# How to Enjoy Calculus

Eli S Pine

Universities Press

How to Enjoy Calculus

**UNIVERSITIES PRESS (INDIA) PRIVATE LIMITED**

*Registered Office*
3-6-747/1/A and 3-6-754/1, Himayatnagar
Hyderabad 500 029 Telangana, India
E-mail: info@universitiespress.com; www.universitiespress.com

*Distributed by*
**Orient Blackswan Private Limited**

*Registered Office*
3-6-752, Himayatnagar, Hyderabad 500 029 Telangana, India

*Other Offices*
Bengaluru, Bhopal, Chennai, Guwahati, Hyderabad, Jaipur,
Kolkata, Lucknow, Mumbai, New Delhi, Noida, Patna, Vijayawada


First Published 2002
Reprinted 2005, 2010, 2012, 2013, 2015, 2017, 2018, 2020

ISBN: 978 81 7371 406 1



*Typeset by*
OSDATA, Hyderabad 500 029

*Printed in India at*
**Yash Printographics.**

*Published by*
© Universities Press (India) Private Limited
3-6-747/1/A and 3-6-754/1,
Himayatnagar, Hyderabad 500 029, Telangana, India

# Contents

# Preface

This book is written for anyone who wants to know what Calculus is all about. Unlike all the other books on this subject, anyone who has knowledge of some algebra and some geometry will be able to understand it. For too long this subject has been mysterious and beyond understanding. With this book the secrets lie exposed at last.

The following persons will be interested in this book:

1. Everyone who has a love for the operation of the human mind and is curious enough to want to know the details of calculus—likely the foremost invention of the human mind.
2. Scientists, engineers and technical people who have studied calculus for years but have never really understood it. Very often, understanding the subject is a luxury that a student cannot afford. As we know, the very, very practical consideration of passing your subjects and finishing school with a degree requires that you pass tests. Consequently, we have memorized our way through this subject. Here's your once-in-a-lifetime chance to overcome that troublesome feeling that you didn't really understand it, although you may even be using it daily now.
3. College students who are about to start the subject. Read this before you go to Math class.
4. High School students who have taken algebra and geometry.
5. Those who have been away from Math for years but who want to get to the meat of the subject as efficiently as possible. The understanding given in this book will enable you to read other books which use Calculus with a gratifying degree of comfort, I guarantee, and your self-confidence will not be shaken easily. Should you feel that something in here could be done more clearly, or if you have any suggestions, they are welcome.

Eli S. Pine

# 1

# Learning Calculus

## The Easiest Way to Learn Calculus

### Calculus renamed

Calculus, first of all, is wrongly named. It should never have been given that name. A far truer and more meaningful name is SLOPE-FINDING, and that is what I shall call it. Slope-finding is really very easy indeed, and is quite beautiful as well. I guarantee that you will learn slope-finding if you have the patience and time to take my hand, as I guide you easily through this subject. In my experience of teaching this subject, no one has gone away without understanding it completely. You may temporarily feel uncomfortable while learning this. You might then put this book down, go out and enjoy yourself, and come back and go into slope-finding again. You may have to do this over and over. Nobody learns easily, and this uncomfortableness is the price everyone must pay while learning.

### Slope-finding

Everything in the world is related in one way or another to something else. For example, let us suppose that you have a job which pays you Rs 6 per hour. Then if you work 10 hours you will have earned Rs 60. If you work 20 hours, your pay is Rs 120, and so on. So, your salary is related to the number of hours you work. Just for easy viewing we make a picture of this earning story like this (Figure 1); now in this figure, it is easy to see that for each hour worked, you get another 6 rupees

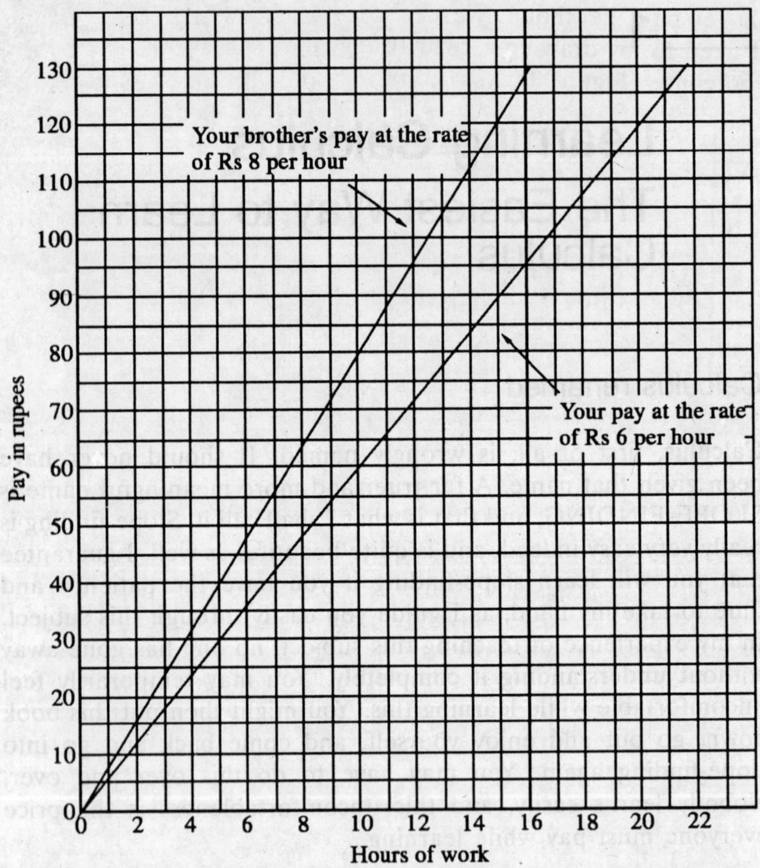


**Figure 1** *Pay resulting from time worked: Notice that a higher rate gives a steeper slope.*

in your paycheck. This picture shows immediately how much you will earn for any number of hours by merely extending the graph as far as you need to. Now suppose your older brother is earning Rs 8 per hour. Notice that the slope of his line is greater (steeper) than yours, and his paycheck will be too, if you both work the same number of hours. So you see, the slope (steepness) of the line is very important; the sharper the slope upwards, the higher your rate of pay, and the more money you will get paid per hour. We can easily measure these slopes if we need to, by drawing a right triangle with the rate line as the hypotenuse (Figure 2).

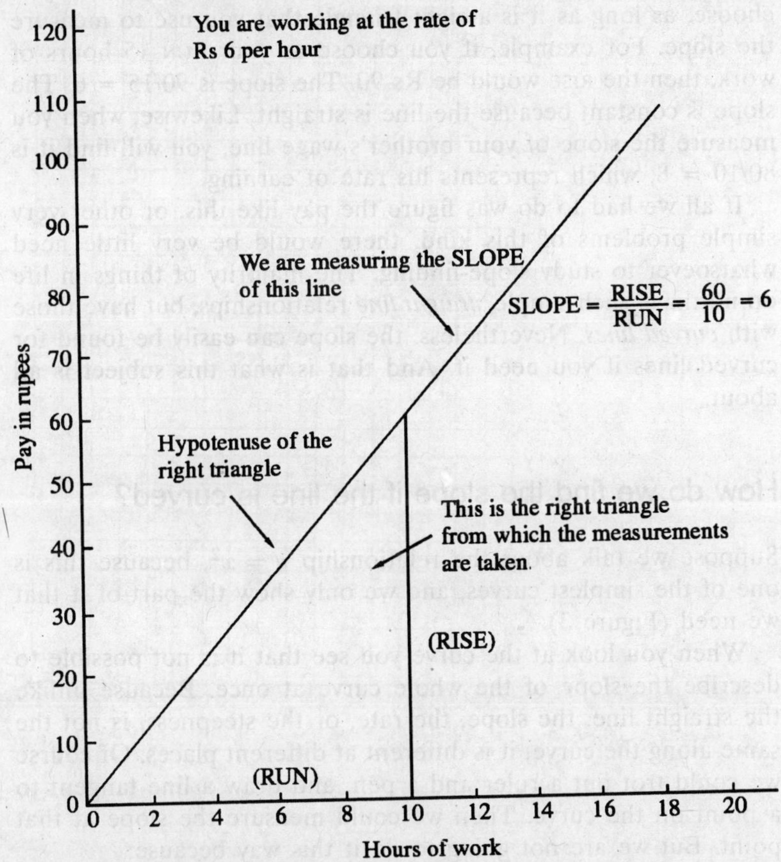


**Figure 2** *The measurement of slope.*

## How slope is measured

Slope is always measured by the length of the vertical leg of this (Figure 2) right triangle (called the RISE), divided by the length of the horizontal leg of the right triangle (called the RUN). The slope of your pay is 6 (which corresponds to your rate of pay, i.e., Rs 6 per hour). Notice that it does not matter how big or little this right triangle is, or which section of the line you choose, as long as it is a right triangle that you use to measure the slope. For example, if you choose as your RUN 15 hours of work, then the RISE would be Rs 90. The slope is $90/15 = 6$. The slope is constant because the line is straight. Likewise, when you measure the slope of your brother's wage line, you will find it is $80/10 = 8$, which represents his rate of earning.

If all we had to do was figure the pay like this, or other very simple problems of this kind, there would be very little need whatsoever to study slope-finding. The majority of things in life do not have such simple *straight line* relationships, but have those with *curved lines*. Nevertheless, the slope can easily be found for curved lines if you need it. And that is what this subject is all about.

## How do we find the slope if the line is curved?

Suppose we talk about the relationship $y = x^2$, because this is one of the simplest curves, and we only show the part of it that we need (Figure 3).

When you look at the curve you see that it is not possible to describe the slope of the whole curve at once. Because unlike the straight line, the slope, the rate, or the steepness, is not the same along the curve; it is different at different places. Of course we could trot out a ruler and a pen, and draw a line tangent to a point on the curve. Then we could measure the slope at that point. But we are not going to do it this way because:

1. It is awkward to do.
2. To measure the slope quickly and accurately with a ruler at a certain point is harder to do than it appears. Try it and see.
3. What we really need is to be able to find the slope anywhere on the curve by a formula of some kind, not by instruments, so that everyone will be able to arrive at the same (correct) value! But is this possible? So come along, let us see where we go!

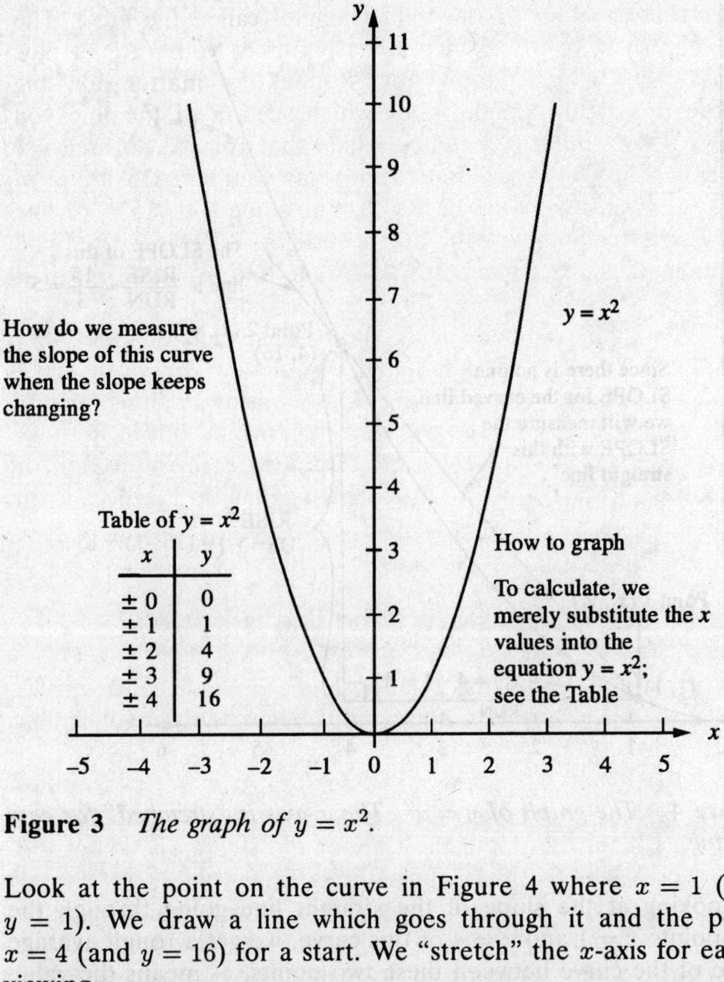


**Figure 3** *The graph of $y = x^2$.*

Look at the point on the curve in Figure 4 where $x = 1$ (and $y = 1$). We draw a line which goes through it and the point $x = 4$ (and $y = 16$) for a start. We "stretch" the $x$-axis for easier viewing.

The slope is always measured by the length of the RISE (height) divided by the length of the RUN of the right triangle shown. In this case,

$$\left.\begin{array}{l}\text{Slope through}\\ \text{points 1 and 2}\end{array}\right\} = \frac{\text{RISE}}{\text{RUN}} = \frac{y_2 - y_1}{x_2 - x_1} = \frac{16 - 1}{4 - 1} = \frac{15}{3} = 5,$$

where $x_1$, $y_1$, is at point 1, and $x_2$, $y_2$ is at point 2.

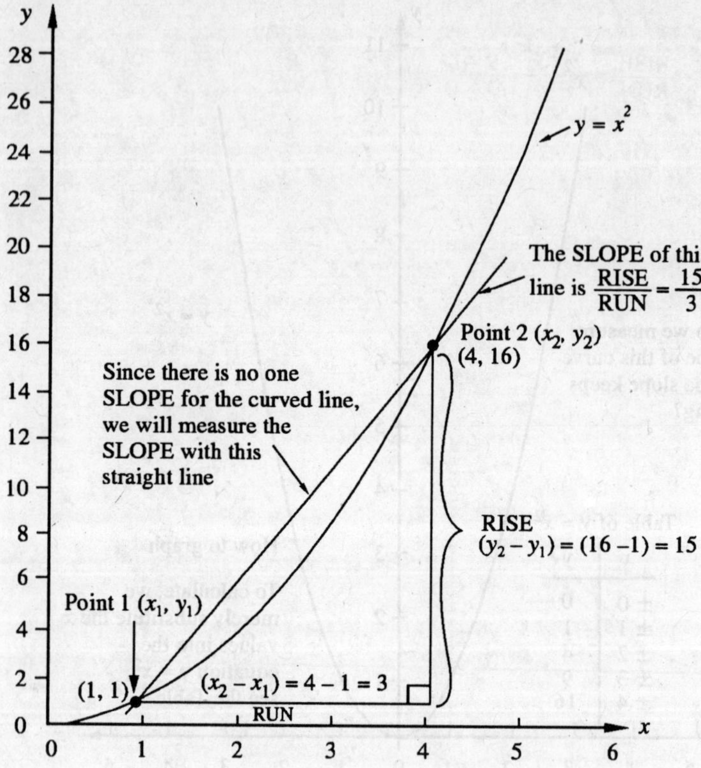


**Figure 4** *The graph of $y = x^2$. The x-axis is "stretched" for easy viewing.*

Looking at the slope of the straight line going through the two points $x = 1$ and $x = 4$ of this curve, we get a rough average slope of the curve between these two points. $y_2$ means the value of $y$ at point 2, $y_1$ means the value of $y$ at point 1, and so on. Notice that the slope 5 does not represent the slope at $x = 1$, nor does it at $x = 4$, nor is it an absolutely exact average slope of the curve between those two points; it is only a "rough average" of this part of the whole curve.

Now let us concentrate on two different and closer points on the same curve (Figure 5), this time at the points on the curve where $x = 1$ and $x = 3$. (The $x_1$, $y_1$ coordinates are the same as

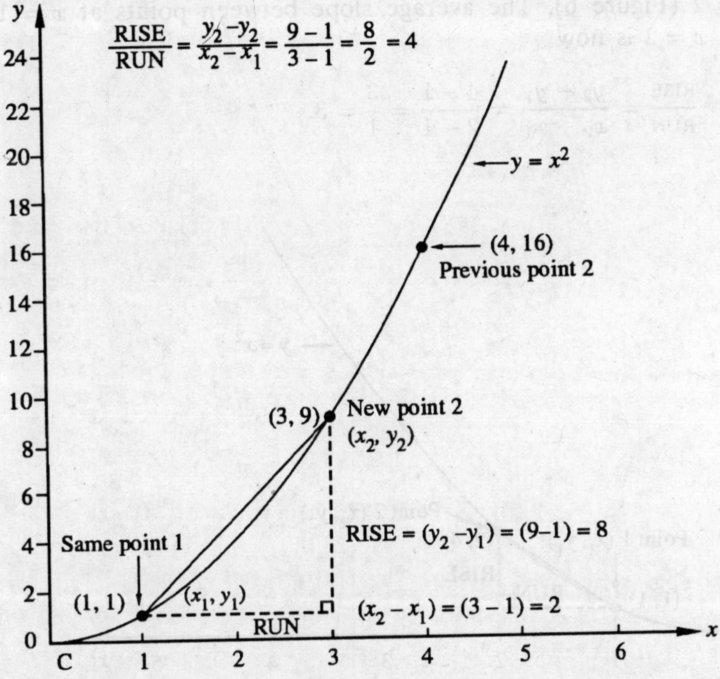


**Figure 5** *Graph of $y = x^2$ with the x-axis "stretched" for easy viewing. New point 2 is now at $x = 3$, $y = 9$ and the new "average" slope of the curve between the points $x = 1$ and $x = 3$ is 4.*

in the last example, but $x_2 = 3$ and $y_2 = 9$). We find the rough "average" slope of the curve between these two points in the same way as before.

$$\left.\begin{array}{l}\text{"Average" slope for}\\ \text{the curve between}\\ x = 1 \text{ and } x = 3\end{array}\right\} = \frac{\text{RISE}}{\text{RUN}} = \frac{y_2 - y_1}{x_2 - x_1} = \frac{9-1}{3-1} = \frac{8}{2} = 4.$$

(the new ''average'' slope)

Now we again stay with the point on the curve at $x = 1$, but we find the "average" slope between the points at $x = 1$ and

$x = 2$ (Figure 6). The average slope between points at $x = 1$ and $x = 2$ is now

$$\frac{\text{RISE}}{\text{RUN}} = \frac{y_2 - y_1}{x_2 - x_1} = \frac{4-1}{2-1} = \frac{3}{1} = 3.$$

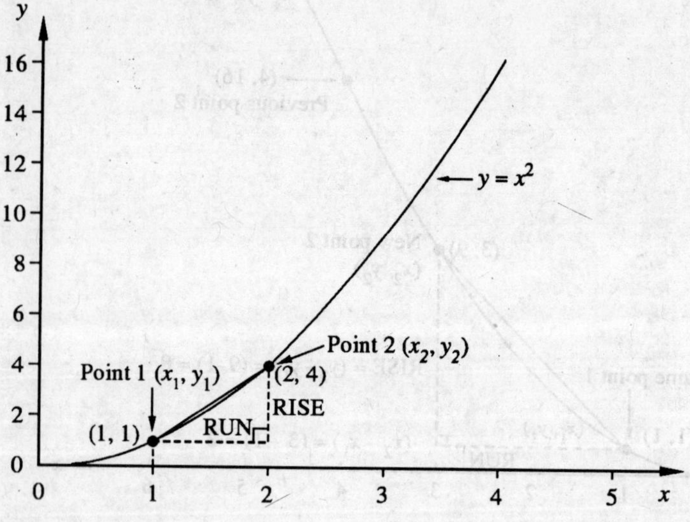


**Figure 6** *Graph of $y = x^2$. The new point 2 is closer at $x = 2$, $y = 4$. The new "average" slope of the curve between $x = 1$ and $x = 2$ is 3.*

The magnitude of the slope, of course, is decreasing, which is not surprising since we are purposely doing this. See if you can guess the purpose as you continue following along. Now we go to two points even closer for the same curve $y = x^2$, where $x = 1$ and $x = 1\frac{1}{2}$. See Figure 7, which has been magnified a little more for easy viewing. The average slope between the points at $x = 1$ and $x = 1\frac{1}{2}$ is now

$$\frac{\text{RISE}}{\text{RUN}} = \frac{y_2 - y_1}{x_2 - x_1} = \frac{9/4 - 1}{3/2 - 1} = \frac{9/4 - 4/4}{3/2 - 2/2} = \frac{5/4}{1/2}$$

$$= \frac{5}{4} \times \frac{2}{1} = \frac{10}{4} = 2\frac{1}{2}.$$

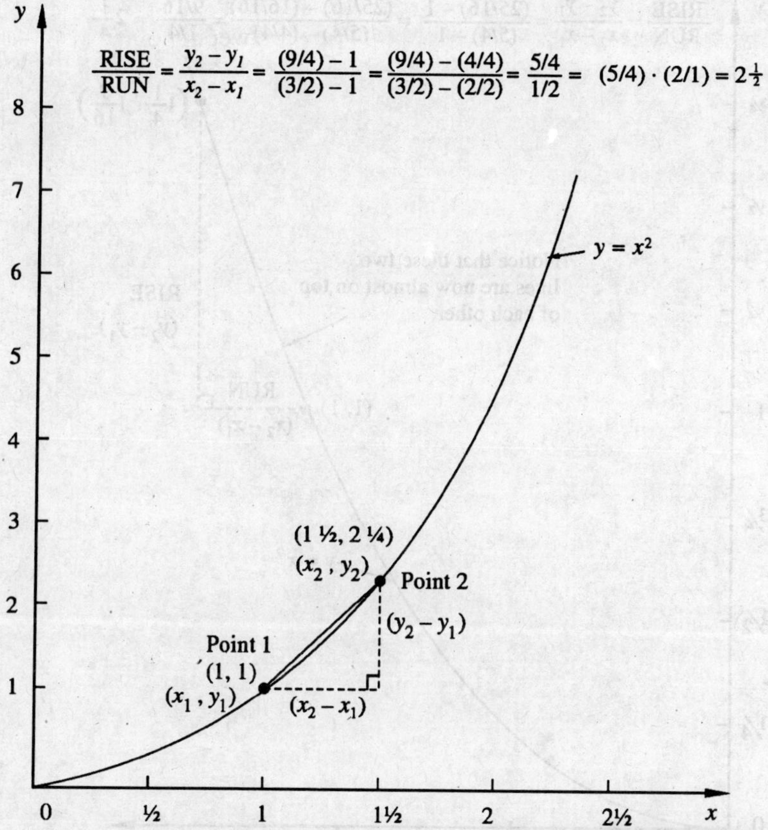


**Figure 7** *Graph of $y = x^2$ with both the $x$ and $y$ axes "expanded" for easier viewing. The "average" slope of the curve between the even closer points $x = 1$ and $x = 1\frac{1}{2}$ is now $2\frac{1}{2}$.*

A very important feature to notice is that the denominator, 1/2, is inverted into 2/1 by virtue of it being a fraction in the denominator, and multiplied with 5/4 to make it $10/4 = 2\frac{1}{2}$.

Now we are going to continue by getting the average slope between two yet closer points, the points on the curve where $x = 1$ and $x = 1\frac{1}{4}$. See Figure 8.

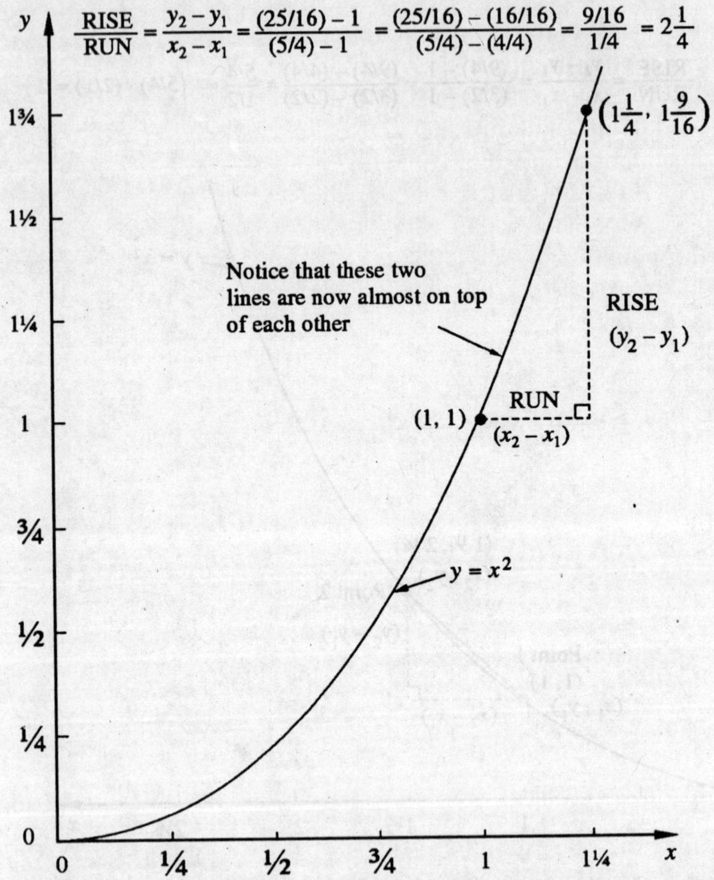


**Figure 8** *Graph of $y = x^2$ expanded even more. The "average" slope of the curve between the points on the curve at $x = 1$ and $x = 1\frac{1}{4}$ is now $2\frac{1}{4}$.*

The average slope between the points on the curve at $x = 1$ and at $x = 1\frac{1}{4}$ is

$$\frac{\text{RISE}}{\text{RUN}} = \frac{y_2 - y_1}{x_2 - x_1} = \frac{25/16 - 1}{5/4 - 1} = \frac{25/16 - 16/16}{5/4 - 4/4} = \frac{9/16}{1/4}$$

$$= \frac{9}{16} \times \frac{4}{1} = 2\frac{1}{4}.$$

(Notice again that the 1/4 in the denominator is inverted to become 4 when multiplying 9/16.)

Since we can easily figure out new average slopes between any two points without the use of a new diagram just with our imagination and the same form of RISE/RUN as before, we continue with points at the curve where $x = 1$ and $x = 1\frac{1}{8}$. The value of $y$ for each value of $x$, again, is found by substituting into the equation $y = x^2$. For $x = 1$, $y = 1$. For $x = 1\frac{1}{8} = 9/8$, $y = (9/8)^2$, or $y = 81/64$, and the average slope between these two points is

$$\frac{\text{RISE}}{\text{RUN}} = \frac{y_2 - y_1}{x_2 - x_1}$$

$$= \frac{81/64 - 1}{1\frac{1}{8} - 1} = \frac{81/64 - 64/64}{9/8 - 8/8} = \frac{17/64}{1/8}$$

$$= \frac{17}{64} \times \frac{8}{1} = \frac{17}{8} = 2\frac{1}{8}.$$

Notice again that the 1/8, when inverted, is 8/1, which is used to multiply the fraction 17/64. Also notice that the denominator always is the simple difference between the two $x$ points, in this case $x = 1\frac{1}{8}$ and $x = 1$, the difference here being 1/8.

Continuing, we now find the average slope for an even smaller difference in length, say, between $x = 1$ and $x = 1\frac{1}{10}$. We determine the corresponding values of $y$ (from $y = x^2$) to be

$$y = (1)^2 = 1 \text{ for point 1,}$$

$$\text{and } y = \left(1\frac{1}{10}\right)^2 = \left(\frac{11}{10}\right)^2 = \frac{121}{100} = 1.21 \text{ for point 2.}$$

Now the "average" slope of the curve between these two points, calculated exactly the same way as before, is

$$\left.\begin{array}{l}\text{average slope between the}\\ \text{points } x = 1 \text{ and } x = 1.1\\ \text{for the function } y = x^2\end{array}\right\} = \frac{\text{RISE}}{\text{RUN}} = \frac{1.21 - 1}{1.1 - 1} = \frac{0.21}{1/10} = 2.1.$$

We notice that, as we find the average slope between $x = 1$ and another point closer and closer to it, as the difference between the $x$ values is getting smaller and smaller, the slope is getting smaller as well and is approaching the magnitude 2.0. It doesn't look like the slope will ever go below 2.0, and it won't simply because at the point on the curve where $x = 1$, the slope is 2 (2 is the limit). Let us see what happens though, as we continue with smaller and smaller distances from $x = 1$, and we can do without the diagrams since they all look about the same. Let us find the average slope between $x = 1$ and $x = 1.01$ (of the function $y = x^2$). When $x = 1.01$, $y = (1.01)^2 = 1.0201$,

$$\left.\begin{array}{l}\text{the average slope}\\ \text{between } x = 1\\ \text{and } x = 1.01 \text{ for the}\\ \text{function } y = x^2\end{array}\right\} = \frac{\text{RISE}}{\text{RUN}} = \frac{1.0201 - 1}{1.01 - 1} = \frac{0.0201}{1/100}$$

$$= 0.0201 \times \frac{100}{1} = 2.01.$$

Notice that in the denominator is the simple difference between the $x$ values making up the "RUN" portion of the slope triangle, 1/100, which becomes inverted and multiplies the smaller number 0.0201, to become $0.0201 \times 100/1 = 2.01$.

We can easily see that the difference (the RISE and the RUN) in the numerator and the denominator is becoming smaller and smaller, and even seems to be in danger of disappearing if it becomes too small. Is this truly the case? I think you are realizing now that the reason why the average slope between two very close points can never disappear altogether is that the denominator, while getting smaller when converted to a fraction, is inverted (for multiplying the numerator), and acts like superman ... It really wallops the numerator with a tremendous kick. Look at this. We look at the points in the same curve $y = x^2$ at $x_1 = 1$ and $x_2 = 1.00001$. Then $y_2 = (1.00001)^2 = 1.0000200001$ and the slope is

$$\left.\begin{array}{l}\text{average slope of } y = x^2\\ \text{between the points } x = 1\\ \text{and } x = 1.00001\end{array}\right\} = \frac{\text{RISE}}{\text{RUN}} = \frac{y_2 - y_1}{x_2 - x_1} = \frac{1.0000200001 - 1.0}{1.0001 - 1.0}$$

$$= \frac{0.0000200001}{\underbrace{0.00001}} = 0.0000200001 \times \frac{100000}{1} = 2.00001.$$

Notice here!

(The denominator is 0.00001, or 1/100,000. When inverted it becomes 100,000.)

This can easily go on and on; with smaller and smaller differences in $x$, the slope will never become less than 2, although it gets closer and closer to 2; so let us try a little device.

## The device (courtesy of Sir Isaac Newton)

Let us try a difference between two points that is so small that instead of between $x = 1$ and $x = 1.000001$, we will go between $x = 1$ and $x = 1.00000000001$, but since we get tired of writing all the numbers after 1, we will write down "very tiny $x$" in their place, that is, we will take "very tiny $x$" to mean 0.00000000001. And we can believe that this is a very, very small $x$ distance indeed, compared to 1. Now instead of finding $y_2 = (1.00000000001)^2$, which is the same as $(1+0.00000000001)^2$, we will find $y_2 = (1 + \text{tiny}\, x)^2$ because it now means exactly the same thing. Now we evaluate $y_2 = (1 + \text{tiny}\, x)^2$ exactly as we did before, and we square $(1 + \text{tiny}\, x)$ just as we have been doing all along with other $x$ values (we shorten "very tiny $x$" even more to "tiny $x$").

Notice the multiplication—if we can multiply symbols, why not words?

$$\begin{array}{r} 1 + \text{tiny}\ x \\ 1 + \text{tiny}\ x \\ \hline 1 + \text{tiny}\ x \qquad\qquad\quad \\ + \text{tiny}\ x + (\text{tiny}\ x)^2 \\ \hline 1 + 2\ \text{tiny}\ x + (\text{tiny}\ x)^2. \end{array}$$

We are getting tired of writing "tiny $x$", so let us choose the symbol $\Delta x$ to mean "tiny $x$". (Now $\Delta x$ does not mean the product of $\Delta$ and $x$. $\Delta x$ is together, and is taken only to mean "tiny $x$".) We now have $y_2 = 1 + 2\Delta x + (\Delta x)^2$. We can remove the parenthesis from $\Delta x$, just leaving $\Delta x$. The third term of course does not mean the $x$ alone is squared, but rather the "tiny $x$" together is squared, that is what is meant here by $\Delta x^2$.

Now calculating the average slope exactly the same way for curve $y = x^2$ between the points $x = 1$ and $x = 1 + \Delta x$, we get

This is $(1+\Delta x)^2$, which is $y_2$ This is $y_1$

$$\left.\begin{array}{l}\text{average slope}\\ \text{between the points } x = 1\\ \text{and } x = 1 + \Delta x \text{ for}\\ \text{the function } y = x^2\end{array}\right\} = \frac{\text{RISE}}{\text{RUN}} = \frac{(1+2\Delta x+\Delta x^2)-1}{(1+\Delta x)-1}$$

This is $x_2$ This is $x_1$

Now we continue exactly in the same way we did before when we used numbers. The ones cancel out in the numerator as well as in the denominator, leaving

$$\left.\begin{array}{l}\text{average slope}\\ \text{between } x = 1\\ \text{and } x = 1 + \Delta x\end{array}\right\} = \frac{\text{RISE}}{\text{RUN}} = \frac{2\Delta x + \Delta x^2}{\Delta x}.$$

Now we separate the terms in the numerator this way: $\frac{2\Delta x}{\Delta x} + \frac{\Delta x^2}{\Delta x}$, which leaves $2 + \Delta x$. Now we do something that may appear very strange! The $\Delta x$ that we have so painstakingly used, we now *cast away* because

1. The $\Delta x$ is very very small *compared with the* 2, and does not operate on the 2 in any substantial way.
2. The $\Delta x$ is no longer in the denominator where it could have been "superman", nor is it being used to *multiply* the 2, which would have been a different story.

We have purposely maneuvered the $\Delta x$ until it got out of the denominator and is by itself. (You see, a number is not nearly as important as where that number is in relation to other numbers. For, the smaller a number is, if it is in the denominator, it carries a wallop when inverted. And it must be inverted to calculate the value of the fraction of which it is a part.)

Now we have found that the average slope between the points at $x = 1$ and $x = 1 + \Delta x$ (for the function $y = x^2$) has the value 2. Of course now you realize that the rough "average" slope we took between the points on the curve at $x = 1$ and $x = 3$ need not bother us much when we take the rough average between $x = 1$ and $x = 1 + \Delta x$ because, since the points are so close together anyway, their average is exact for all purposes. If you have understood all this, then you will have little trouble going all the rest of the way. Let us think about this average slope. Imagine in your mind that the smaller the distance between the points, the closer to the value of 2 we will get, since the only part we have thrown away, the $\Delta x$ at the final step, will be so small as to be insignificant compared with the magnitude of the original slope of 2. And if the distance between the two points is so very small, the two points practically blur together for any purpose at all, practical or theoretical.

## Finding the slope at another point on $y/x^2$

Now we can find the slope at any point of the function $y = x^2$ in the same way. For example, let us find the slope at $x = 3$. We find the average slope between the two points on the curve at $x = 3$ and at $x = 3 + \Delta x$ — a "very tiny" distance, $\Delta x$, along the curve. Then exactly as before, $y^2 = (3 + \Delta x)^2$. Squaring the term $3 + \Delta x$,

$$\begin{array}{r} 3 + \Delta x \\ 3 + \Delta x \\ \hline 9 + 3\Delta x \phantom{+ \Delta x^2} \\ + 3\Delta x + \Delta x^2 \\ \hline 9 + 6\Delta x + \Delta x^2 \end{array}.$$

Then the average slope between $x = 3$ and $x = 3 + \Delta x$ for the function $y = x^2$

$$= \frac{\text{RISE}}{\text{RUN}} = \frac{y_2 - y_1}{x_2 - x_1} = \frac{(9 + 6\Delta x + \Delta x^2) - 9}{(3 + \Delta x) - 3}$$

$$= \frac{6\Delta x + \Delta x^2}{\Delta x} = \frac{6\Delta x}{\Delta x} + \frac{\Delta x^2}{\Delta x} = 6 + \Delta x.$$

And we discard the $\Delta x$ since, as before, compared to the magnitude of 6, the $\Delta x$ is so very tiny, and is being added to the 6 anyway, not divided or multiplied. Do you still ask, why do we then need the $\Delta x$ in $3 + \Delta x$ and use it, yet drop it at the end from the $6 + \Delta x$? This is for the same reason we gave before, that the $\Delta x$ as part of the number 6 is of negligible importance, since their comparative magnitude is so very very far apart. But when we use the $\Delta x$ *to establish the rate* as in the $(3 + \Delta x)^2$, then the position of the $\Delta x$ is, towards that establishment, of very great importance. I'm sure you can think of another good reason as well.

There is a purported Chinese proverb which fits this situation very well. It goes something like this. A mighty river starts with only a trickle of water, whose course it is easy to change but only while it is still at its source. When the trickle of water has become part of a vast river which meets the sea, changing the direction of that trickle of water at the later location has not the slightest effect on the river. Correspondingly, the error that we get in dropping the $\Delta x$ at the end of the process of finding the slope is so small as to be almost incalculably small. Now we know that the slope is 4 for the same function $y = x^2$ at the point $x = 2$ (really the average between the two points $x = 2$, and $x = 2 + \Delta x$). If we find the slope at various other values of $x$ we get

for $x = 1$, the slope is 2,
for $x = 2$, the slope is 4,
for $x = 4$, the slope is 8,
for $x = 25$, the slope is 50.

Whatever point we choose, whose slope is desired, we can automatically get the slope in the case of $y = x^2$ by multiplying the $x$ value desired by 2. *This means that we have a general formula for finding the slope at any point for the function* $y = x^2$, and all we need to do is multiply by 2 any $x$ value desired. The general[1] formula is, for the general slope for all points, $2x$ (for

[1] General does not mean the average over the entire curve (which is useless). It does mean a formula that will give *the slope* at any point of the curve, just as the original function gives *the coordinates* at any point on the curve.

the function $y = x^2$). This of course means that we need only substitute for the $x$ in the above (slope $= 2x$) to automatically get the slope. For example if $x = 4$, then slope $= 2x = 2(4) = 8$, so of course $2x$ is the *general slope* formula for all points of the function $y = x^2$. Now aside from guessing that the slope is always two times the value of $x$ at any point of the function $y = x^2$, we need only take one further step to see how we could otherwise arrive at the same conclusion.

Instead of taking a particular point to look at and find the slope, such as $x = 1$, or $x = 3$, or $x = 4$, we will not take any particular value of $x$. We leave the $x$ alone, so as to stand for *any* number we may wish, without assigning any particular value to $x$. Then, if we start with the general value $x$, and go a small distance $\Delta x$ from it, then for the same function $y = x^2$ we have (to calculate the RISE)

$$y_2 = (x + \Delta x)^2 \quad \text{or} \quad \begin{array}{r} x + \Delta x \\ x + \Delta x \\ \hline x^2 + x\Delta x \qquad \\ + x\Delta x + \Delta x^2 \\ \hline x^2 + 2x\Delta x + \Delta x^2. \end{array}$$

Last time we used a number in place of $x$ (not for the $x$ in $\Delta x$)

Now as before, the slope is still measured in exactly the same way:

$$\left.\begin{array}{l}\text{Slope of the} \\ \text{function } y = x^2 \\ \text{between the points} \\ x \text{ and } x + \Delta x\end{array}\right\} = \frac{\overbrace{x^2 + 2x\Delta x + \Delta x^2}^{\text{This is } y_2} - \overbrace{x^2}^{\text{This is } y_1}}{\underbrace{x + \Delta x}_{\text{This is } x_2} - \underbrace{x}_{\text{This is } x_1}} = \frac{\text{RISE}}{\text{RUN}}.$$

(Notice that the $x^2$ we are subtracting from the numerator is the original function from the equation $y = x^2$, also that the denominator always ends up to be $\Delta x$ alone.)

$$= \frac{2x\Delta x + \Delta x^2}{\Delta x} = 2x + \Delta x = 2x.$$

We have dropped the $\Delta x$ again, for the same reason as before. (The $\Delta x$ is not in the denominator, nor is it multiplied by anything; it is merely to be added to the $2x$). Thus we have a general formula for the slope (which is $2x$) for *any* value of $x$, not just a particular value of $x$. Now we will define the RISE, that is $y_2 - y_1$, which is also very small; we will call it $\Delta y$ (which is the vertical side of the right triangle; we cannot call it $\Delta x$, because it is not the same length. $\Delta x$ is the horizontal side of the right triangle) and $x_2 - x_1$, the RUN, is $\Delta x$. Then $\Delta y/\Delta x$ is the slope, and is most accurate when the $\Delta x$ is so very very tiny that it is almost zero! This is equivalent to dropping $\Delta x$ at the end. $\Delta x$ is made to "approach zero". But it should not be zero itself, since dividing by a very very small number, even one that is so small as to be very nearly zero is perfectly legal.[2] By universal agreement the process (symbol) of slope-finding is written as

$$\frac{\Delta y}{\Delta x} = \boxed{\frac{dy}{dx}},$$ ⟵ The general slope

when $\Delta x$ is so small as to nearly approach zero, or, $\Delta x \to 0$.

The other ways in which $(dy/dx)$ is denoted are: $f'(x), y', D_x y$, $d(y)/dx$, and $\dot{y}$. And for the function $y = x^2$, the general slope for any point on it is

$$\frac{\Delta y}{\Delta x} = \frac{dy}{dx} = 2x.$$
(when $\Delta x \to 0$)

The arrow $\to$ does not mean 'becomes zero', it means becomes so small as to be almost zero

So, by universal agreement we say that $dy/dx$ is the notation for the general *slope* for the curve $y = x^2$, or $dy/dx = 2x$ for the function $y = x^2$. (Notice that $dy$ is *not* the product of $d$ and $y$, nor is $dx$ the product of $d$ and $x$. You cannot separate the $x$ from $dx$, nor the $y$ from $dy$.)

And that is what we have been doing, slope-finding, and we don't even have to draw the function $y = x^2$ anymore to find the slope at any point, since all we need is the $x$ value which we directly substitute into $dy/dx = 2x$.

---

[2] See Appendix B, which shows why division by zero is absolutely not allowed in any mathematics.

With this little device of finding the average slope between two very close points of a function, we now have the advantages which may not seem important to you as yet. The advantages so far are these:

1. We can describe the general slope of all points of a function in a very concise and simple manner, just as the function itself is described, by a FORMULA.
2. We do not need to draw or measure a function in order to find the slope or rate, at any point.
3. We can find the slope at any point of a function very quickly.

Maybe these will somehow lead to other advantages unforeseen now?

## Finding the general slope for yet another function

Let us now go on to another function. Let us try to find the general slope at all points for the function ($y = x^2 + x + 1$), and we will make firstly this table for graphing. See Figure 9.

Let us look at any general value of $x$ for this new function ($y = x^2 + x + 1$). As before, we mainly need to find $y_2$, which is obtained by substituting $x + \Delta x$ where $x$ was before; then $y_2 = (x + \Delta x)^2 + (x + \Delta x) + 1$. Multiplying the first term, and gathering all the like terms together,

$$\begin{array}{l} x + \Delta x \\ x + \Delta x \\ \hline x^2 + x\Delta x \\ \quad\;\; + x\Delta x + \Delta x^2 \\ \hline x^2 + 2x\Delta x + \Delta x^2. \end{array}$$

And now,

$$y_2 = (x^2 + 2x\Delta x + \Delta x^2) + (x + \Delta x) + 1.$$

The general slope formula is

$$\frac{\text{RISE}}{\text{RUN}} = \frac{\Delta y}{\Delta x} = \frac{y_2 - y_1}{x_2 - x_1}$$

$$= \frac{\overbrace{(x^2 + 2x\Delta x + \Delta x^2) + (x + \Delta x) + 1}^{y_2} - \overbrace{(x^2 + x + 1)}^{y_1}}{\Delta x}.$$

This is $x_2 - x_1$ This is the original function $y = x^2 + x + 1$

Gathering the terms together,

$$\frac{\Delta y}{\Delta x} = \frac{dy}{dx} = \frac{2x\Delta x + \Delta x^2 + \Delta x}{\Delta x}$$

when $\Delta x \to 0$

$$= 2x + \Delta x + 1$$

$$= 2x + 1.$$ We have cast out only the $\Delta x$ term, as before

$(dy/dx)$ is also called the *first derivative* besides the *general slope.*

It is clear that we have obtained the general slope for any value of $x$ for the function $y = x^2 + x + 1$, and this is $dy/dx = 2x + 1$.

If we want to know the slope of *any* function, not only of $y = x^2$ and $y = x^2 + x + 1$, the effort would be much greater—for instance, the function $y = x^3 + 4x^2$, not to mention more complicated functions. Notice that we would have to cube $x + \Delta x$, since $x + \Delta x$ must be cubed to find $y_2$. Fortunately, there is a short cut that is so simple it is dramatic. Here it is: To find the general slope for all equations of the form $y = x^n$, we need only follow this simple recipe, or manipulation ($n$ is the exponent):

$$\frac{dy}{dx} = nx^{n-1}.$$ This is the short cut

And we do this for each term separately if the equation has other terms. The following examples and others that follow later, will show the kinds of manipulations that you should practice doing.

**Question: Can the same method as used for the function $y = x^2$, be used for this function ($y = x^2 + x + 1$) for finding the slope?**

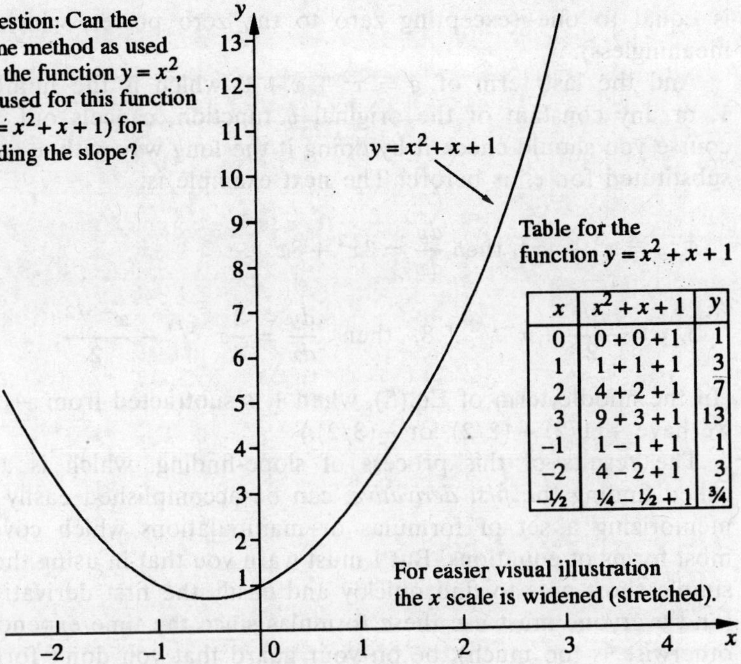


**Table for the function $y = x^2 + x + 1$**

| $x$ | $x^2 + x + 1$ | $y$ |
|---|---|---|
| 0 | 0 + 0 + 1 | 1 |
| 1 | 1 + 1 + 1 | 3 |
| 2 | 4 + 2 + 1 | 7 |
| 3 | 9 + 3 + 1 | 13 |
| –1 | 1 – 1 + 1 | 1 |
| –2 | 4 – 2 + 1 | 3 |
| –½ | ¼ – ½ + 1 | ¾ |

**For easier visual illustration the $x$ scale is widened (stretched).**

**Figure 9** *Graph of* $y = x^2 + x + 1$.

For example,

1. $y = x^2$, then $\dfrac{dy}{dx} = 2x$.

2. $y = x^2 + x + 1$, then $\dfrac{dy}{dx} = 2x + 1$.

3. $y = 3x^4 - 2x^2 + \dfrac{x}{2} - 6$, then $\dfrac{dy}{dx} = 12x^3 - 4x + \dfrac{1}{2}$.

Note the middle term $x$ of equation 2. $x$ can be said to be $x^1$, or $x$ to the first power. A letter, or number with no exponent is always taken to be to the first power. Now we multiply the coefficient of $x$ (which is 1) and subtract 1 from the exponent. This leaves $1x^0$, and $x^0 = 1$, since anything to the zero power is equal to one (excepting zero to the zero power, which is meaningless).

And the last term of $y = x^2 + x + 1$, which is the number 1, or any constant of the original $y$ function, cancels out. Of course you should check it by doing it the long way with $x + \Delta x$ substituted for $x$, as before. The next example is:

4. $y = x^3 + 4x^2$, then $\dfrac{dy}{dx} = 3x^2 + 8x$.

5. $y = \dfrac{x^{3/4}}{2} + x^{-1/2} + 8$, then $\dfrac{dy}{dx} = \dfrac{3}{8}x^{-1/4} - \dfrac{x^{-3/2}}{2}$.

(In the middle term of Eq.(5), when 1 is subtracted from $-1/2$, we have $-(1/2) - (2/2)$, or $-(3/2)$.)

The results of this process of slope-finding, which is also called finding the *first derivative*, can be accomplished easily by memorizing a set of formulas or manipulations which covers most forms of equations. But I must warn you that in using these simple short cuts to find quickly and easily the first derivatives (and everyone must use these formulas since the time expended otherwise is too much), be on your guard that you don't forget why we use the short cut and how it was done. For example, for the function $y = x^3 + 2x + 1$, when we look for the general slope, we get $dy/dx = 3x^2 + 2$. The transformation looks like magic. It looks like it's unmathematical. In fact it looks unreal. The change to the first derivative, using the short cut form, believe it or not, is such a radical change to the mind that it looks like it's against all logic, and different from the other math you have learned. Consequently, the logical process in your mind undergoes the trauma of *shock* since using the short cut doesn't look valid. You will, therefore, tend to forget algebra and maybe even ordinary arithmetic now and then if you don't watch yourself. It is almost like temporary amnesia. But if you think about the whole thing over and over again till it is perfectly clear in your mind, you will have no more trouble with it. Also, slope-finding insists that there be no errors in your work because in ordinary arithmetic, where adding and subtracting are most common, an error may be of small consequence, but in slope-finding, where you are dealing with division by very small numbers and multiplication of very large numbers, even a small error brings huge errors in the end.

I will show you, by example, how I use a small number to get a very large error with large consequences. Let us use the number 2. If your rate of pay is Rs 6.00 per hour, and you are mistakenly paid for 38 hours of work instead of 40, then there is an error of 2, that is, 2 hours of work. You will have been underpaid by Rs 12.00. But let us say that your employers make a different kind of error with the number 2. Let us say they mistakenly compute your pay at Rs 4.00 per hour instead of Rs 6.00 per hour. This also is an error with the number 2. But the consequences of this error are far more hurtful because your paycheck will be Rs 240.00. You suffer a Rs 80.00 hole in your paycheck by this error, the result of which also has been caused by the number 2. So you see, in slope-finding you are also dealing with the same type of error. Therefore, be extra careful in your arithmetic and your algebra. And notice that the Rs 2.00 in the rate of pay is very, very important, but in your Rs 240.00 paycheck the addition of Rs 2.00 more or less is almost insignificant. And do not allow the traumatic shock of differentiation (finding the first derivative, or slope-finding), cause you to make silly mistakes in arithmetic, even though it seems that the rules of algebra and common sense are broken, which they certainly are not. Remember that differentiating does not change all those logical ways which you have previously learned, your arithmetic, algebra, etc. They are exactly as valid and intact as they were before. All you are doing when you memorize this short cut for finding the slope, is finding the slope without having to do all the work, as we have gone through each time. It is not the slope-finding that is hard, it is guarding against silly arithmetic errors, believe it or not.

## Other forms of equations

For the simple equations mentioned before, like $y = x^2$, we have $dy/dx = 2x$. The form of the equation $y = x^2$, is $y = x^n$. So then $dy/dx = nx^{n-1}$. And for each different form of equation, to find the general slope, there is another recipe. For instance suppose we have the function $y = (2x)/(1 - x^2)$. We could find $dy/dx$, the general slope, doing it the long way, like the other examples which we have done before, or we can memorize this general formula:

$$\frac{d}{dx}\left(\frac{u}{v}\right) = \frac{v\frac{du}{dx} - u\frac{dv}{dx}}{v^2},$$

called the quotient form.

Here if $u = 2x$ and $v = 1 - x^2$, then $du/dx = 2$, and $dv/dx = -2x$, we substitute these in the formula for slope:

$$\frac{dy}{dx} = \frac{(1 - x^2)(2) - (2x)(-2x)}{(1 - x^2)^2} \quad \texttt{Multiplying out the numerator}$$

$$= \frac{2 - 2x^2 + 4x^2}{(1 - x^2)^2} \quad \texttt{Gathering the terms together in the numerator}$$

$$= \frac{2x^2 + 2}{(1 - x^2)^2}.$$

This, then, is the general slope for all points of the function $y = 2x/(1 - x^2)$.

Please check it by graphing the function, then by using the $dy/dx$ equation to compute the slope at several points, and finally, also verifying these answers by actually measuring the slope with a ruler. The other simple forms for finding the general slope to equations (the first derivatives) are given in Table 1.1 (these, and many more, can be found in any calculus text).

Given below are the "cook book" formulae which describe a procedure, or a set of instructions, and should be understood in the following way (see Form 3 in Table 1.1).

$$\frac{d}{dx}(u + v - w) = \frac{du}{dx} + \frac{dv}{dx} - \frac{dw}{dx}.$$

If we wish to differentiate a function of the form $y = u + v - w$, where $u$, $v$, $w$ represent terms, then we do this to them

$$\frac{dy}{dx} = \frac{du}{dx} + \frac{dv}{dx} - \frac{dw}{dx}. \quad \texttt{Differentiate term by term}$$

**Table 1.1** *Table for the slope-finding short cut*[3] *(these are the most-used forms).*

| | | |
|---|---|---|
| 1. | $\dfrac{d(c)}{dx} = 0$ | where $c$ is any number or constant. |
| 2. | $\dfrac{d(x)}{dx} = 1$ | |
| 3. | $\dfrac{d}{dx}(u + v - w) = \dfrac{du}{dx} + \dfrac{dv}{dx} - \dfrac{dw}{dx}$ | |
| 4. | $\dfrac{d}{dx}(cv) = c\dfrac{dv}{dx}$ | |
| 5. | $\dfrac{d}{dx}(uv) = u\dfrac{dv}{dx} + v\dfrac{du}{dx}$ | |
| 6. | $\dfrac{d}{dx}(v^n) = nv^{n-1}\dfrac{dv}{dx}$ | Here $v$ stands for more than one term, all to the power $n$. |
| 7. | $\dfrac{d}{dx}(x^n) = nx^{n-1}$ | This form is what we are using mostly. |
| 8. | $\dfrac{d(u)}{dx(v)} = \dfrac{v\frac{du}{dx} - u\frac{dv}{dx}}{v^2}$ | |
| 9. | $\dfrac{d(u)}{dx(c)} = \dfrac{du/dx}{c}$ | |
| 10. | $\dfrac{dy}{dx} = \dfrac{dy}{dv} \cdot \dfrac{dv}{dx}$ | $y$ being a function of $v$. |
| 11. | $\dfrac{dy}{dx} = \dfrac{1}{dx/dy}$ | $y$ being a function of $x$. |

Now that we know how to find the general slope for all points of an equation, let us practice a little.

### Example 1

Given the function $y = x^2$, what is the slope at the point $x = 5$?

### Solution

We see that $y = x^2$ is of the form $y = x^n$, and $dy/dx = nx^{n-1}$, which, for the case $y = x^2$, is $dy/dx = 2x$, and for the point $x = 5$ of the curve, merely[4] substitute 5 for $x$; $dy/dx = 2(5) = 10$. Check this by actually graphing $y = x^2$ and measuring the slope at $x = 5$. (Most errors in this type of problem are in substituting 5 in the equation $y = x^2$ instead of in $dy/dx = 2x$.)

---

[3] Appendix D shows a typical proof of these formulae.

[4] You will recall the "$x$" in "$dx$", is not to be taken for, or equivalent to, the "$x$" in "$2x$". The $dx$, though using the letter "$x$", cannot be separated.

EXAMPLE 2

For the function $y = 2x^3 - 3x + 7$, what is the slope at $x = 2$?

SOLUTION

We get it by simply differentiating each term separately, $dy/dx = 6x^2 - 3$ (see Form 3 in Table 1.1).

Now to find the slope at the point $x = 2$, we substitute 2 for $x$ (do not substitute 2 for $x$ in the original equation $y = 2x^3 - 3x + 7$, since this will not give the slope; only the $dy/dx$ equation will);

$$\frac{dy}{dx} = 6(2)^2 - 3 = 24 - 3 = 21.$$

The slope at the point $x = 2$, for the equation $y = 2x^3 - 3x + 7$ is 21.

Check this result by graphing the function $y = 2x^3 - 3x + 7$, then measuring the slope at $x = 2$. (See how much more difficult it is to find the slope by measuring it with a ruler? We also must accurately graph it even before we can do that!)

EXAMPLE 3

Find $dy/dx$ of the function $y = (a^2 - x^2)^{1/2}$.

SOLUTION

Utilizing Form 6 of Table 1.1, $\dfrac{d(v^n)}{dx} = nv^{n-1}\dfrac{dv}{dx}$,

$$\frac{dy}{dx} = \tfrac{1}{2}(a^2 - x^2)^{-\frac{1}{2}}(-2x).$$

Notice the exponent $\left(\frac{1}{2}\right) - 1 = -\left(\frac{1}{2}\right)$; also $\left(\frac{d}{dx}\right)(-x^2) = -2x$

(do not ignore the negative sign of the term $-x^2$). Continuing and simplifying,

$$\frac{dy}{dx} = -\frac{x}{(a^2 - x^2)^{1/2}}. \qquad \texttt{Answer}$$

EXAMPLE 4
From the function $y = x^2$, locate the coordinates at the point on this curve where the slope is 5.

SOLUTION
We know that $dy/dx = 2x$ since $dy/dx$ is the general slope for all points on the curve $y = x^2$. Since our example requires that the slope should be equal to 5, substitute 5 for $dy/dx$, thereby getting $5 = 2x$; hence $x = 5/2$ (where $dy/dx$, the slope, is 5). And from the original equation $y = x^2$, it is very easy to obtain the $y$ coordinate at $x = 5/2$. We merely substitute 5/2 for $x$ (in $y = x^2$) obtaining $y = 25/4$. (A frequent error is to substitute 5/2 into the $dy/dx$ equation.)

Thus the coordinates have been found:

$$x = 5/2 \quad \text{and} \quad y = 25/4. \qquad \texttt{Answer}$$

Please check it yourself by graphing very accurately $y = x^2$ between $x = 2$ and $x = 3$.

EXAMPLE 5
Find the slope at $x = 1$ of the function $y = x^4 + 1$.

SOLUTION
We find the first derivative, which will give all the slopes automatically for the function $y = x^4 + 1$. This is $dy/dx = 4x^3$, where $dy/dx$ is *the slope*. For the slope at the particular point $x = 1$, substitute for $x$ into $dy/dx = 4x^3$, obtaining

$$\frac{dy}{dx} = 4(1)^3 = 4. \qquad \texttt{Answer}$$

Check by graphing.

EXAMPLE 6
Given $y = x^5 + 3x^4 - 1$, find the first derivative.

$$\frac{dy}{dx} = 5x^4 + 12x^3 \text{ (Forms 1 and 3 of Table 1.1).} \qquad \texttt{Answer}$$

EXAMPLE 7

Given $y = 3 - x^2 - x/2$, find the slope at $x = -2$.

Slope $= 3\frac{1}{2}$. Answer

EXAMPLE 8

Given the circle $x^2+y^2 = 4$, find the coordinates where the slope is 1.

SOLUTION

$x^2 + y^2 = 4$

$y^2 = 4 - x^2$

$y = \pm(4 - x^2)^{1/2}$.

First change into the form $y$ = etc., because we must get $y$ alone to the first power on the left side of the equation

Now differentiate to find the general slopes for all points of the curve.

$$\frac{dy}{dx} = \pm\frac{1}{2}(4 - x^2)^{-1/2}\underbrace{(-2x)}$$

If you omitted this part, the problem is wrong (see Form 6 in Table 1.1)

$$1 = \pm\frac{1}{2}(4 - x^2)^{-1/2}(-2x)$$

Now substitute $dy/dx = 1$

$$1 = \frac{-x}{\pm(4 - x^2)^{1/2}}$$

$\pm(4 - x^2)^{1/2} = -x$

$4 - x^2 = x^2$

$4 = 2x^2$

$2 = x^2$

$x \cong \pm 1.41\ldots$

We square both sides of the equation because we can do anything to one side if we also do it to the other side

The sign $\cong$ means here "approximately". See Figure 10. Notice that $x = \pm 1.41$ is correct as per the diagram, and that even the negative distance $x = -1.41$ makes sense here and is needed.

Are there any other examples that would help fix firmly in your mind that $dy/dx$ is really *the slope*, and only *the slope*? You can experiment now with your own functions by accurately graphing them and measuring the slopes at various points of your own choice.

Appendix F contains a computer program that will evaluate numerically the slope of any function. For example, what is the slope at $x = 5$ for the function $y = x^2$?

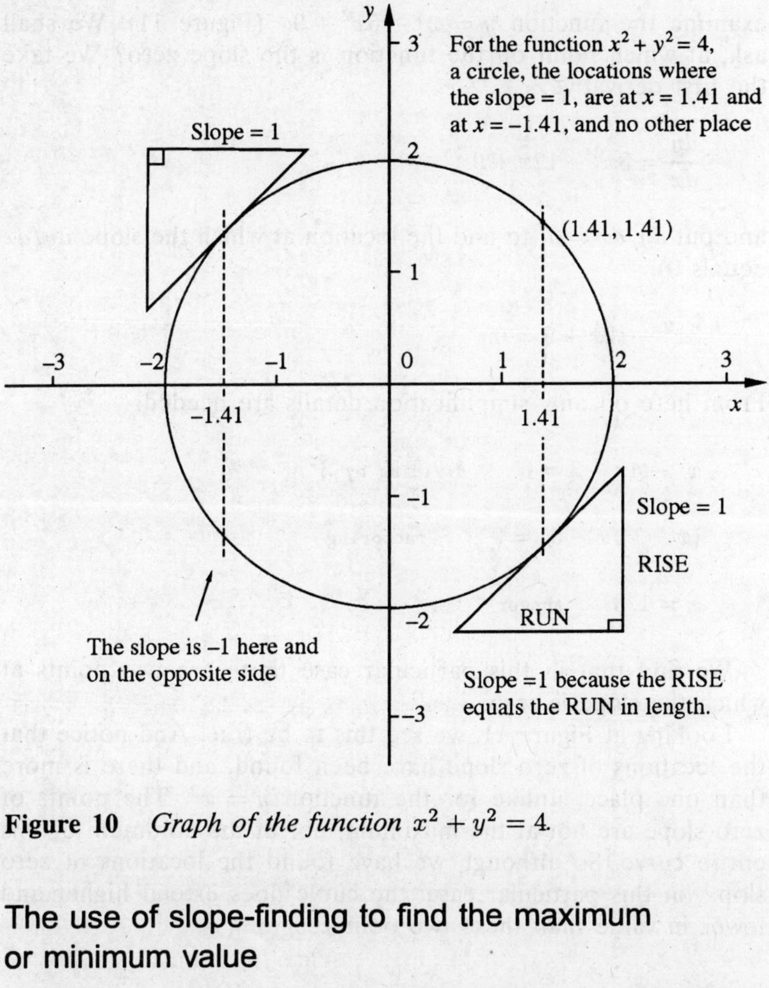


**Figure 10** *Graph of the function* $x^2 + y^2 = 4$.

## The use of slope-finding to find the maximum or minimum value

Now that we can find the slope for an equation, here comes the first use to which it will be put. If we find the first derivative of the function $y = x^2$, which is $dy/dx = 2x$, and we wish to know the point at which the slope is zero, it remains for us only to substitute 0 for $dy/dx$, because $dy/dx$ is the slope! We then obtain $0 = 2x$ or $x = 0$. Figure 3 shows this to be true. *Zero slope* is the point at which the function $y = x^2$ reaches its lowest, and therefore the minimum value. Let us likewise examine the function $y = x^3 - 6x^2 + 9x$ (Figure 11). We shall ask, at which point on the function is the slope zero? We take the first derivative

$$\frac{dy}{dx} = 3x^2 - 12x + 9$$

and put $dy/dx = 0$ (to find the location at which the slope $dy/dx$ equals 0),

$$3x^2 - 12x + 9 = 0.$$

From here on only simplification details are needed.

$$x^2 - 4x + 3 = 0 \qquad \texttt{dividing by 3}$$

$$(x - 1)(x - 3) = 0 \qquad \texttt{factoring}$$

$$x = 1, 3. \qquad \texttt{Answer}$$

We find that in this particular case there are two points at which the slope is zero.

Looking at Figure 11, we see this to be true. And notice that the locations of zero slope have been found, and there is more than one place, unlike for the function $y = x^2$. The points of zero slope are not at the maximum, nor at the minimum of this entire curve. So although we have found the locations of zero slope, in this particular case, the curve does extend higher and lower in value than these two points.

## Some word problems using slope-finding

Here is how you convert a word problem into a curve, and then put into use the *zero slope*. Later we won't need the curve.

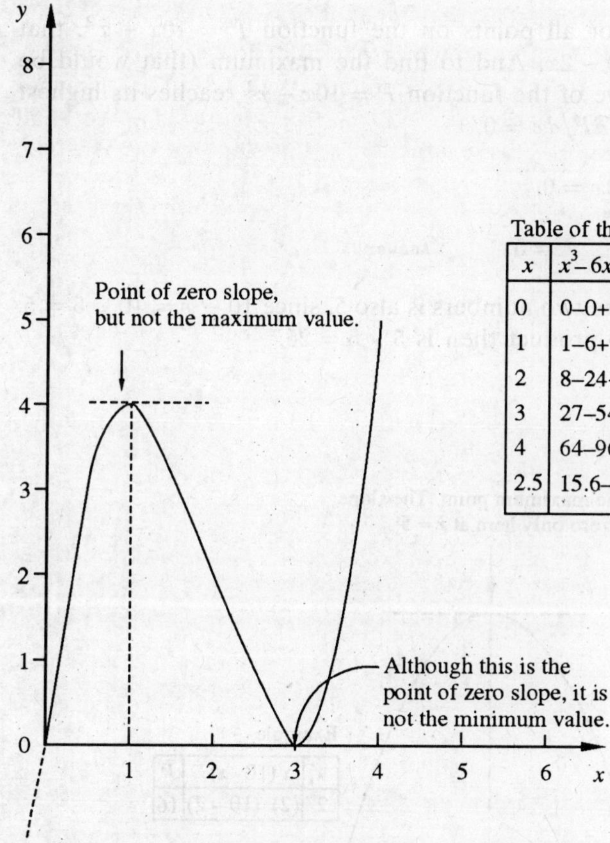


Table of the function

| $x$ | $x^3 - 6x^2 + 9x$ | $y$ |
|---|---|---|
| 0 | 0–0+0 | 0 |
| 1 | 1–6+9 | 4 |
| 2 | 8–24+18 | 2 |
| 3 | 27–54+27 | 0 |
| 4 | 64–96+36 | 5 |
| 2.5 | 15.6– etc. | 0.625 |

**Figure 11** *Graph of the function* $y = x^3 - 6x^2 + 9x$.

EXAMPLE 9

Find two positive numbers whose sum is ten, and whose product is a maximum (Figure 12).

SOLUTION

If we let one number be $x$, the other number must be $10 - x$, since the sum of the two numbers $(10 - x) + x$ is equal to 10. Now their product is $P = x(10 - x)$, and multiplying, we get $P = 10x - x^2$.

We take the first derivative which will give us the equation of the slope for all points on the function $P = 10x - x^2$, that is, $dP/dx = 10 - 2x$. And to find the maximum (that would be where the curve of the function $P = 10x - x^2$ reaches its highest point), we set $dP/dx = 0$.

Then $10 - 2x = 0$.

$2x = 10; \quad x = 5.$ Answer

The other of the two numbers is also 5, since $10 - x = 10 - 5 = 5$. The maximum product then is $5 \times 5 = 25$.

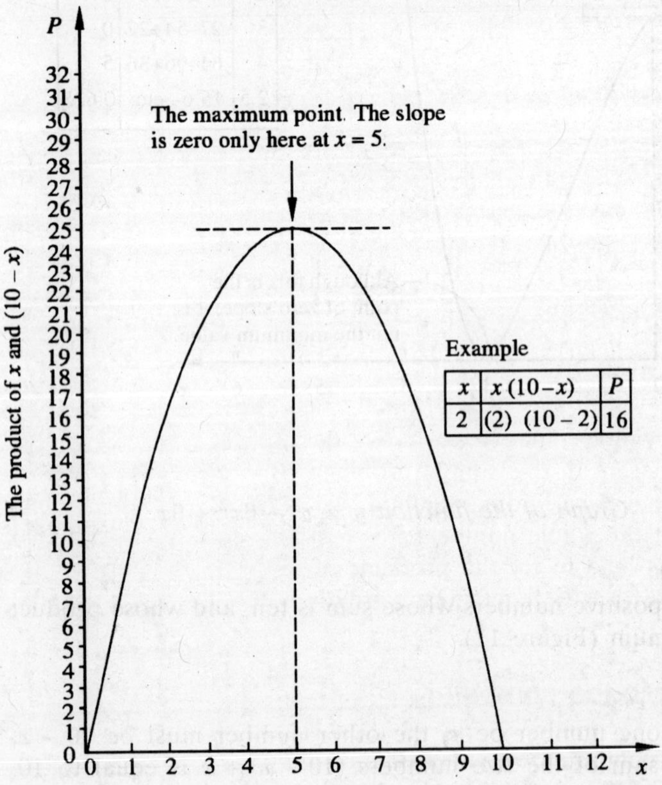


**Figure 12** *Graph of $x$ vs. $P$; $P = x(10 - x)$, where $x$ is one number, $(10 - x)$ the other, and $P$ is their product.*

Check this result with the graph (Figure 12). There are other things to notice in Figure 12. Notice that, first of all, the answer $P = 25$ and $x = 5$ is the correct one, that the maximum value is where the slope is zero (at the top). Notice that the second letter in the numerator of $dP/dx$ (here $P$) always refers to the vertical axis, likewise the second letter in the denominator (here $x$) always refers to the horizontal axis. We really do not need slope-finding in this case to do this problem. It could have been done by trial and error. For we would try all numbers whose sum is 10, and multiply them together, as in Table 1.2.

**Table 1.2**

| $x$ | | $10 - x$ | | Product |
|---|---|---|---|---|
| 0 | × | 10 | = | 0 |
| 1 | × | 9 | = | 9 |
| 2 | × | 8 | = | 16 |
| 3 | × | 7 | = | 21 |
| 4 | × | 6 | = | 24 |
| 5 | × | 5 | = | 25 ← Note that this is the maximum |
| 6 | × | 4 | = | 24 |
| 7 | × | 3 | = | 21 |
| 8 | × | 2 | = | 16 |
| 9 | × | 1 | = | 9 |
| 10 | × | 0 | = | 0 |

The advantage that *slope-finding* has in this type of example is not apparent in this simple problem. But notice, anyway, that without the slope-finding method, we would have had to also verify that the whole number, 5 × 5, is the maximum, here. We would have had to try the products of non-whole numbers, like 4.9 × 5.1 and 4.95 × 5.05, etc., greatly adding to the labor.

## Bringing in the computer

With the computer we can examine hundreds, even thousands of other combinations very quickly. If you do not wish to use the computer, you can ignore these examples. We can use the following program (in BASIC):

```
10   x = 0            Starting value of x
20   P = 10 * x - x ∧ 2   The equation P = 10x - x^2
                      ( in BASIC)
30   PRINT x, P       Print P corresponding to x
40   x = x + .1       Tells what increment (.1) to
                      add onto x. This is not an
                      algebraic equation, but merely
                      instructs the computer.
                      There is a difference.
50   GOTO 20          Return to Step 20 for yet
                      another calculation
```

In seconds, watching the computer terminal, you can see that $P = 25$ is not exceeded. You can stop the computer as soon as you see that. Try changing line 40 to $x = x + .01$, which will go to hundredths. Still $P = 25$ cannot be exceeded. And so on. By letting the derivative equal to zero, we have found the extreme value of this curve.

This was a simple problem that did not require slope-finding. Suppose we had asked this one:

EXAMPLE 10

Find three positive numbers (not necessarily whole numbers), one of which is twice that of another, and the sum of the three numbers is 100, such that the product of all three numbers is a maximum. We can see that the problem now has become very complicated, and trial and error is now very difficult (except by computer), though still possible. Compare this problem, where an additional complication requires the use of slope-finding to arrive at a solution, with the problem shown in Figure 17, whose additional complication makes finding the solution by slope-finding virtually impossible. However, the computer can be used.

SOLUTION (BY SLOPE-FINDING)

Let $x$ = one number, then $2x$ = the second number, and $100 - 2x - x$ = the third number. Then the product, $P$, of all 3 numbers is $P = (x)(2x)(100 - 3x)$. Multiply them;

$$P = 200x^2 - 6x^3. \left\{ \begin{array}{l} \texttt{Notice that the thing we wish to find,} \\ P\texttt{, is alone on the left-hand side of the} \\ \texttt{equation. The only other unknown, } x\texttt{,} \\ \texttt{is on the right-hand side of the equation} \end{array} \right.$$

We can now find the first derivative, so that we find the location of zero slope(s), which will then allow us to find the maximum value of all the various products.

$$\frac{dP}{dx} = 400x - 18x^2.$$

We let $dP/dx = 0$, to find the location where the slope is zero. And from here on, the remaining work is only details (yet careful details).

$$400x - 18x^2 = 0$$

$$200x - 9x^2 = 0 \quad \texttt{Divide by 2}$$

$$x(200 - 9x) = 0 \quad \texttt{Factoring}$$

One solution is $x = 0$, and the other is obtained from

$$200 - 9x = 0, \text{ i.e., } x = 200/9 = 22.222...$$

Remember that the locations of zero slope gotten by use of the first derivative, are *not necessarily* the locations of maxima or minima (but are necessarily the *only locations* of zero slope) when we investigate the curve of the original equation. A rough sketch of this equation shows that the locations of zero slope are indeed exactly at $x = 0$ and $x = 22.222...$, and that the rest of the curve has either negative values of $x$ or smaller values of $P$, both of which are not wanted. We are also assured that the curve has no other surprising turns because there are no other locations of zero slope.

If $x = 22.222$ (rounded out), then $2x = 44.444$ and $100 - 3x = 33.333$. Their product is $32,922$ (rounded off). There are no other 3 numbers whose product exceeds this. (This is what is meant by the maximum of the products.) We can now realize the tremendous advantage that slope-finding has for this problem, since getting the answer was possible by trial and error, but with an awful lot of labor. Of course if we solve the problem by trial and error, then we do not have to make an equation for the problem. But this is not the right way, and as we continue I think you will see that there are very great advantages to making an equation for a problem, but certainly, slope-finding cannot be used unless there is an equation to use it on. Nor can the computer be of use without the correct equation. While the computer can do many problems, by no means can it do all calculus problems.

## The language of nature is the language of extreme

There is a special meaning to finding the minimum value. There is

> "... a tendency in nature to work with the greatest possible ease or the least possible effort. A ray of light reaching the eye from an object seen in a mirror has minimized its route by striking and leaving the mirror at equal angles. Two soap bubbles clinging together adjust so as to have the least possible surface area consistent with their content."
>
> —*Mathematics*, by David Bergami, Time Life Books.

Knowing this fact has been a great help in solving all kinds of scientific and engineering problems by making use of the "principle of least work" and the "principle of least energy", etc. Some actual engineering illustrations (though they are not easy to follow) can be found on page 258, *Theory of Elasticity*, S P Timoshenko and J N Goodier. These, and other similar principles, can also be found in nearly every scientific text. So you see, finding the extreme minimum is not at all just an exercise in enjoyable mathematics, it is also a very practical thing.

## Area problems

Suppose a rectangular area of 3,000 square feet is to be enclosed by a fence, then divided down the center by another fence. If the length of fence in the North–South direction (represented by $y$) costs Rs 1 per running foot, and the length of fence in the East–West direction (represented by $x$) costs Rs 2 per running foot, find the dimensions of $x$ and $y$ that would result in a fence of minimum cost. Find that cost (Figure 13).

As in all problems in slope-finding, or calculus, nothing can be done unless the problem is first made into an equation.

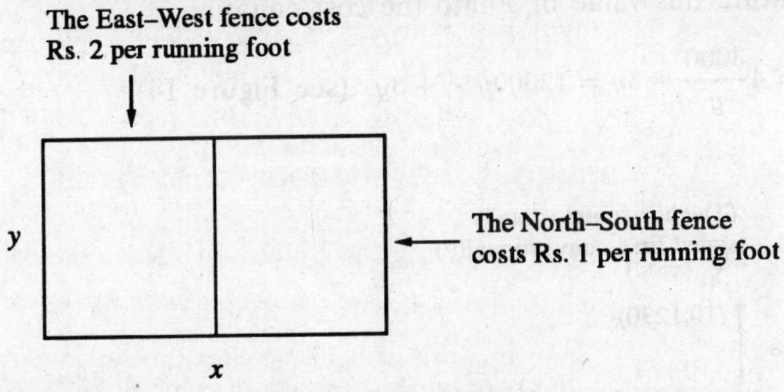


**Figure 13** *Rectangular areas enclosed by a fence.*

We must construct an equation that gives the problem in terms of cost $C$, and in terms of only one other variable. Now let $x$ equal the length of the plot of land in the $x$-direction, and let $y$ equal the length of the plot of land in the $y$-direction.

For the entire rectangle, Area $A = x \cdot y$. This area has been given to be $3,000$ square feet, then $x \cdot y = 3000$.

The total cost involves multiplying the length of one side by the cost per unit length, then adding together the cost of all the sides. For example, the cost of one side in the $x$-direction would be $2x$, because $x$ is the number of feet and 2 is the cost in rupees per foot. Adding the individual cost of each side, we obtain

$$C = 2x + 2x + y + y + y = 4x + 3y.$$

Now we have described the cost situation in terms of two unknowns, $x$ and $y$. We must further describe either $x$ in terms of $y$, or $y$ in terms of $x$, so that we end up with cost in terms of only one unknown. (In a graph, each axis stands for one thing only; in this case $C$ would be the vertical axis, and either $x$ or $y$ would be the horizontal axis).

Solving for $x$,

$$x = \frac{3000}{y}.$$

We substitute this value of $x$ into the cost equation:

$$C = 4\frac{3000}{y} + 3y = 12000y^{-1} + 3y \quad \text{(see Figure 14).}$$

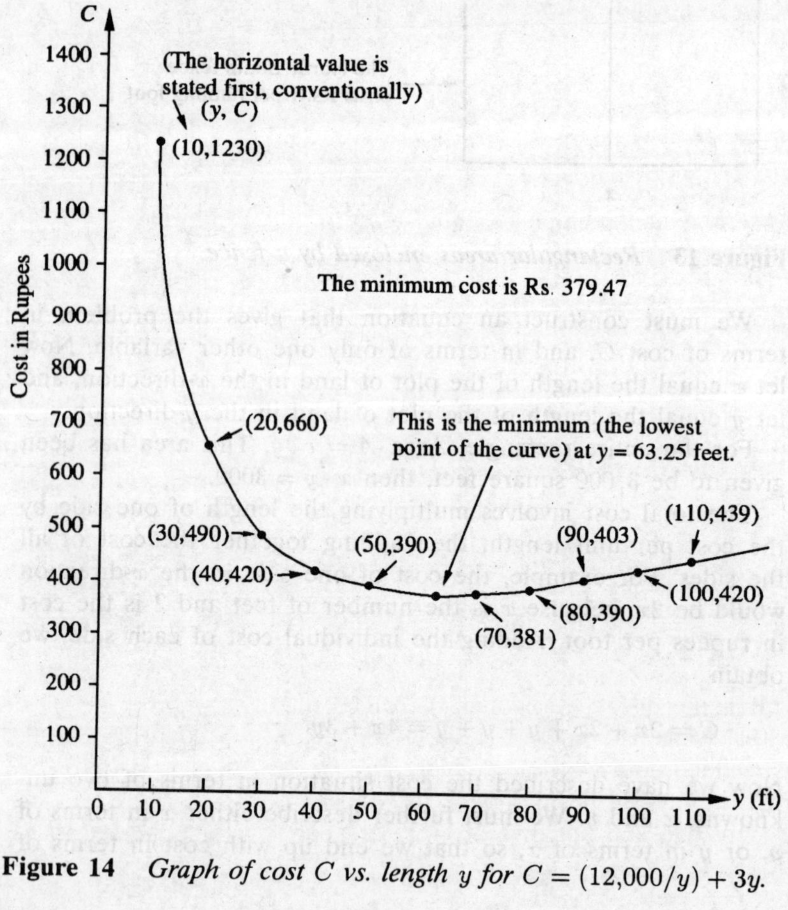


**Figure 14** *Graph of cost $C$ vs. length $y$ for $C = (12{,}000/y) + 3y$.*

Now, and only now, are we in a position to find the slope at all points on this curve, remembering that $C$ (the left-hand side of the equation) will be on the vertical axis, and $y$ will now be the horizontal axis. It will be a convenience and a general rule to *always* use the unknown on the left-hand side of the equation (the dependent variable $C$ in this case) as the vertical axis, and the independent variable on the right-hand side of the equation ($y$ in this case) for the horizontal axis. The graph of such a function will always be one ($C$) in terms of another ($y$), and the derivative will always be the slope at any point on this curve in terms of these two things, and nothing else. Taking the derivative of cost with respect to $y$ (and we could just as easily have found the derivative of cost with respect to $x$),

$$\frac{dC}{dy} = -12,000y^{-2} + 3.$$

Since we are only interested in the location of the minimum cost (not any other cost), and that cost will be located at the zero slope, we let the slope be equal to zero, and we replace $dC/dy$ with zero and solve.

$$0 = -12{,}000y^{-2} + 3$$

$$12{,}000y^{-2} = 3$$

$$\frac{12{,}000}{y^2} = \frac{3}{1} \quad \texttt{Cross multiply}$$

$$3y^2 = 12,000$$

$$y^2 = 4000$$

$$y = \pm 63.25 \text{ feet.}$$

This is the distance at which there is a minimum cost. (See the graph) Since the negative answer for length is of no interest to us (indeed you must judge every answer for physical meaning), we use only the positive part, $y = 63.25$ (in this particular case, the positive and negative answers are equal in magnitude) and substitute it into any previous equation that will enable us to get closer to our answer. In this case

$$x = 3000/y$$

and substituting the value of $y$,

$$x = 3000/63.25, \text{ i.e., } x = 47.43 \text{ ft.}$$

Those then are the dimensions necessary to obtain the minimum cost. That minimum cost is found from the cost equation:

$$C = \frac{4(3000)}{y} + 3y.$$

Substituting for $y$,

$$C = \frac{4(3000)}{63.25} + 3(63.25).$$

The minimum cost is $C$ = Rs 379.47. **Answer**

Notice, we find the location where the slope is zero, then we determine at what magnitude the slope is equal to zero? The magnitude is "read off" of the vertical axis (ordinate axis). True, we don't need slope-finding if we graph the problem, because we could just read off the answer from the graph. Slope-finding avoids the necessity of graphing the problem, but far more importantly, near the end of this book you will see how entire equations are born, a thing that no amount of graphing or computers will do! By experimenting yourself, you will see that no matter what values of $x$ and $y$ that you try, as long as you meet the conditions mentioned in the problem, there will be no cost that will be less than the cost we have just found. (You can use the same computer program shown previously to check this out. Use the cost equation of this problem.) This then, is the minimum value. Finding the minimum value, or finding the maximum value is also known as "extrema" problems (finding the extreme value). Here are some more problems that deal with areas. These problems are completely worked out in Appendix A.

PROBLEM 1

The government has to fence in 60,000 square feet of land in a rectangular plot along a straight highway. The fence costs Re 1 per foot along the highway and Rs 0.75 per foot along the other three sides. How much of each fence is necessary for minimum cost? What is this minimum cost? What are the dimensions?

PROBLEM 2

A farmer wants to fence in 200,000 square feet of land in a rectangular plot, then divide that plot into three equal plots. Find the least amount of fence necessary to do this.

PROBLEM 3

Suppose there is an isosceles triangle, the two sides are equal and are 10 ft. in length. You want to know what height the triangle should be such that the area of the triangle is a maximum. Also, find this area. Notice (Figure 15) that if we bring the sides almost together, it will result in a very small area. And if the sides are very wide apart also, a very small area results. But at a certain intermediate distance apart, the area of the triangle will be an extreme maximum ("extrema").

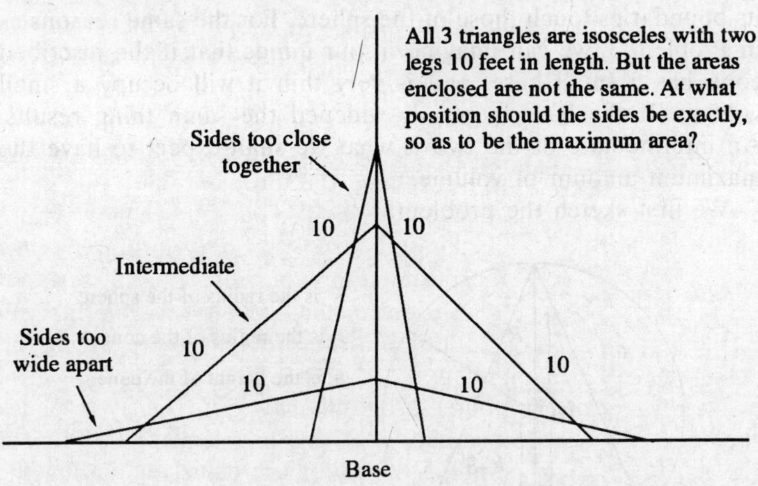


**Figure 15** *Isosceles triangles (two sides equal in length). Exactly how should the legs extend so that the triangle has a maximum area?*

PROBLEM 4
Given a semi-circle, whose equation is $x^2 + y^2 = r^2$, find the dimensions of the rectangle with the largest possible area that can be inscribed in it, and find that area. After calculating the answer to this problem, graph the area of various inscribed rectangles. See if your answer checks with the result which you got by the method of slope-finding! You can see that this problem can also be done by drawing several of these rectangles, and by trial-and-error find which has the maximum area. Do the problem both ways. (Solution in Appendix A).

## Volume problems

Of course the same procedure used in the area problems applies exactly to volume problems. Suppose we look at this one:

PROBLEM 5
Given a sphere, find the dimensions of an inscribed cone, such that the cone will occupy the maximum amount of volume within the sphere. An inscribed cone means it is fitted tightly such that its boundaries touch those of the sphere. For the same reasons as in Problem 3, we can imagine in our minds that if the inscribed cone has a small base, and is very thin it will occupy a small volume. If the base is greatly widened the same thing results. An intermediate of the two is what we shall expect to have the maximum amount of volume.

We first sketch the problem:

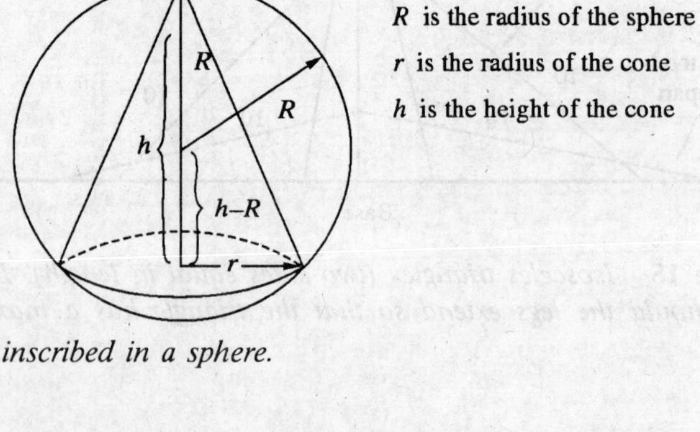


*Cone inscribed in a sphere.*

We must specify the thing we are to maximize, which is the cone's volume. The volume of a cone is (you would look it up in a reference book)

$$V = \frac{1}{3}\pi r^2 h.$$

Now, as you can see from the sketch, there are two variables, $h$ and $r$ ($R$ is a constant). We must eliminate either $h$ or $r$, in order to differentiate (find the slope) for this volume formula! No matter how many unknowns are on the right hand side of the equation, if we can (with the proper relationships) change them into one unknown, we can solve it.

Notice that the Pythagorean formula can be used (it is very commonly used in all manner of problems);

$$(h - R)^2 + r^2 = R^2.$$

The hardest part of slope-finding is the algebra involved in finding the needed relationships. Expanding this equation so as to solve for either $r$ or $h$, we get

$$h^2 - 2hR + R^2 + r^2 = R^2, \quad \text{or} \quad h^2 - 2hR + r^2 = 0.$$

(Remember that $R$ is a constant.)

We could solve for $h$, but we notice that it is far easier to solve for $r^2$, which is, after all, just as good:

$$r^2 = 2hR - h^2$$

which we substitute into $V = (1/3)\pi r^2 h$ to get

$$V = \frac{\pi h(2hR - h^2)}{3}$$

$$= \frac{2\pi Rh^2}{3} - \frac{\pi h^3}{3}.$$

We are now ready to differentiate (slope-find), since we have what we are looking for to maximize (the volume), $V$ alone on the left-hand side of the equation and only one variable, $h$, on the right-hand side. Of course, we could draw the curve for this last equation using $V$ for the ordinate (the $y$-axis) and $h$ for the abscissa ($x$-axis). The top of this curve would give us the maximum value of $V$ without the need for slope-finding. But we will go ahead with this method, without drawing the curve of $V$ vs. $h$,

$$\frac{dV}{dh} = \frac{4}{3}\pi Rh - \pi h^2$$

We now let $\dfrac{dV}{dh} = 0$, and so

$$\pi h^2 = \frac{4}{3}\pi Rh.$$

Simplifying further by dividing both sides by $\pi h$;

$$h = \frac{4}{3}R.$$

This is the answer, and we could also solve for $r$ if we wish, and for volume as well. Solving for $r$ in the equation

$$h^2 - 2hR + r^2 = 0;$$

by substituting the value of $h = \dfrac{4}{3}R$,

$$\left(\frac{4}{3}R\right)^2 - 2\left(\frac{4}{3}R\right)R + r^2 = 0.$$

Then simplifying further,

$$\frac{16}{9}R^2 - \frac{24}{9}R^2 + r^2 = 0,$$

and $r^2 = (8/9)R^2$, then $r = \pm(2/3)R\sqrt{2}$. (We ignore the negative root.)

For the volume of the cone, we use the volume formula $V = \frac{1}{3}\pi r^2 h$. Substituting for $r^2$ and $h$:

$$V = \frac{\pi}{3}\underbrace{\left(\frac{8}{9}R^2\right)}_{\text{(for } r^2\text{)}}\underbrace{\left(\frac{4}{3}R\right)}_{\text{(for } h\text{)}}.$$

$$V_{\text{cone}} = R^3\pi\frac{32}{81}.$$

Just for our interest, we compare this volume with that of the total volume of the sphere; we get

$$V_{\text{sphere}} = \frac{4}{3}\pi R^3 \;,$$

$$\frac{V_{\text{cone}}}{V_{\text{sphere}}} = \frac{\frac{32}{81}\pi R^3}{\frac{4}{3}\pi R^3} \;,$$

$$\frac{V_{\text{cone}}}{V_{\text{sphere}}} = \frac{32}{81}\cdot\frac{3}{4} = \frac{8}{27} \;,$$

or, that this cone occupies about 30% of the total volume of the sphere. First we flex our calculus muscles, later we'll look for practicality. Notice that there is no way possible to do this problem without calculus, and this is only a representative of a multitude.

Try this one yourself.

PROBLEM 6

Given a sphere, find the dimensions of an inscribed cylinder which occupies as much volume as possible (a maximum of volume). Also, find this volume and compare it with that of the sphere (see Appendix A for the solution).

## Problems with distance

Suppose there are 2 towns, let us call them town A and town B (Figure 16). Now town A is 3 miles east of a diner on the main highway, which runs north-south. And 5 miles south of the diner, 12 miles east of the highway, is town B. A water tower is to be built by the highway, and it is desirable that the pipe from towns A and B be such that the total amount of pipe length be a minimum. Where along the highway should the water tower be built? In a real-life situation, it is certain that other considerations would enter into the problem, such as natural obstacles, mountains, or streams. However, there is no reason why these cannot just as the easily be included into the problem, as long as the situation can be put into mathematical terms. Here, then, is the picture:

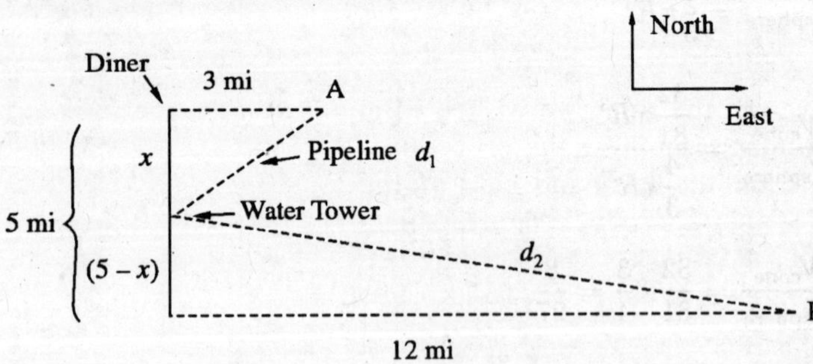


**Figure 16** *Water for two towns.*

Since we do not yet know where to install the water tower along the highway, we should name either the top distance to the diner $x$, or the bottom distance. In any case, the other distance must be $5 - x$, since when the two parts $x$ and $5 - x$ are added together, we get the total of 5. Now, it is our task to describe the length of the pipeline in terms of only one unknown. Call the total length of the pipeline, $P$. We notice that two right triangles are formed. Call the length of the pipeline from Town A to the water tower as $d_1$ and the other pipeline distance $d_2$. Now $P = d_1 + d_2$. We can easily find a relationship by the theorem of Pythagoras. For $d_1$ it is

$d_1^2 = x^2 + (3)^2$, `Isn't the theorem of Pythagoras used very often?`

and $d_1 = \pm\sqrt{x^2 + 9}$. `Ignore the negative length`

We do the same for $d_2$:

$$\begin{aligned} d_2^2 &= (5 - x)^2 + (12)^2, \\ &= x^2 - 10x + 25 + 144, \\ &= x^2 - 10x + 169, \\ d_2 &= \sqrt{x^2 - 10x + 169}. \end{aligned}$$

Then, as we said, $P = d_1 + d_2$, and substituting for $d_1$ and $d_2$,

$$P = (x^2 + 9)^{1/2} + (x^2 - 10x + 169)^{1/2}.$$

We have got the thing we wished to minimize (or, if we wanted to maximize something the procedure is identical). We have got $P$ alone on the left-hand side of the equation, and only one variable, $x$, on the right. We are ready to find the slope. By using the differentiating formula dealing with this exponential type of equation, as mentioned in Table 1.1, we get

$$\frac{dP}{dx} = \frac{1}{2}(x^2 + 9)^{-1/2}(2x) + \frac{1}{2}(x^2 - 10x + 169)^{-1/2}(2x - 10).$$

Now we let $\dfrac{dP}{dx} = 0$, since we want the location of $x$ at zero slope.

$$0 = \frac{1}{2}(x^2 + 9)^{-1/2}(2x) + \frac{1}{2}(x^2 - 10x + 169)^{-1/2}(2x - 10).$$

A little simplification yields

$$-\frac{x}{(x^2 + 9)^{1/2}} = \frac{x - 5}{(x^2 - 10x + 169)^{1/2}}.$$

The easiest way to deal with this situation is to square both sides of the equation. (You can do anything you want to one side of an equation, as long as you do the same thing to the other side). Notice, in squaring, the negative sign disappears:

$$\frac{x^2}{x^2+9} = \frac{(x-5)^2}{x^2-10x+169}.$$

It may be simpler at this point to invert both sides of the equation. Now you know that slope-finding is the easiest part of the problem. It is the setting up of the problem and the algebra of working it through that is much harder (not to mention avoiding silly little errors of arithmetic).

$$\frac{x^2+9}{x^2} = \frac{x^2-10x+169}{(x-5)^2}.$$

If we separate the numerator of the right-hand side of the equation, we can "see the light".... (This trick is sometimes used.)

$$\frac{x^2+9}{x^2} = \frac{(x^2-10x+25)+144}{(x-5)^2}.$$

$$\frac{x^2}{x^2} + \frac{9}{x^2} = \frac{(x-5)^2}{(x-5)^2} + \frac{144}{(x-5)^2}.$$

$$1 + \frac{9}{x^2} = 1 + \frac{144}{(x-5)^2}.$$

Subtracting one from each side,

$$\frac{9}{x^2} = \frac{144}{(x-5)^2}.$$

Take the square root of both sides:

$$\pm\frac{3}{x} = \pm\frac{12}{(x-5)}$$

Both sides are plus and minus. Since multiplying by like signs gives a positive product and multiplying by unlike signs a negative product, there are only two combinations and not four

$$\frac{3}{x} = \frac{12}{x-5}$$ `First use plus on both sides (''like'' signs)`

$$3x - 15 = 12x$$

$$9x = -15; \quad x = -\frac{5}{3}.$$

This answer is negative. This cannot be because there is no negative length involved. So ignore this answer. Try the other combination, i.e., one side is negative, the other positive ("unlike" signs):

$$-\frac{3}{x} = \frac{12}{x-5}$$

$$12x = -3x + 15$$

$$15x = 15; \quad x = 1.$$

This is the answer! It is only a matter of chance that it is a whole number. The other two combinations of signs will give the same answers again. Thus, the water tower should be built $x = 1$ mile (south of the diner) to achieve a pipeline to towns A and B in which the sum of the pipeline lengths will be a minimum. You should check this answer on graph paper. You will find that no other value uses less pipeline from the water tower to towns A and B. The combined distance is $P = (x^2+9)^{1/2} + (x^2 - 10x + 169)^{1/2}$, where we substitute $x = 1$. The minimum length of pipeline is $P = (10)^{1/2} + (160)^{1/2} = 15.81$ miles.

## The computer

Use the computer to see if the answer $x = 1$ really does give the smallest value of $P$ (pipe length).

This is the equation for length:

$$P = (x^2 + 9)^{1/2} + (x^2 - 10x + 169)^{1/2}.$$

We write this program (BASIC language), which can compute any value of $x$ you desire.

PROGRAM 1

```
10   x=.5                       Tells the computer where to start.
20   P=SQR(x∧2+9) +             The equation for pipe length
     SQR(x∧2-10*x+169)          that we found.
30   PRINT x, P                 Print out the values
                                of x and P
40   x=x+.1                     This gives the increment, or
                                how much of an increase in x
50   IF x>2 GOTO 100            Tells the computer
                                when to quit.
60   GOTO 20                    Return for another calculation
                                of P vs x
100  END                        Stops the computer
```

```
Answers
  x        P
  .5     15.8573   ←   Notice the decreasing values of P
  .6     15.8406
  .7     15.8277
  .8     15.8186
  .9     15.8131
 1.0     15.8113   ←   Smallest value of P
 1.1     15.8131
 1.2     15.8183   ←   Notice the increasing values of P
etc.
```

No matter what values of $x$ that you search through, you will be unable to find a smaller value for $P$ than when $x = 1$. Try these changes, which narrows the search. Here we are evaluating between $x = .9$ and $x = 1.1$, with increments of 0.01.

```
10  x = .9               The new starting x value
40  x = x + .01          The new increment
50  IF x > 1.1 GOTO 100  Tells the program when
                         to stop searching
```

Try any values you can think of. See Appendix C for more computer searching.

## Should slope-finding be used all the time?

Obviously, the same principal of the previous problem can be applied to any number of towns, and for any number of different situations. (Compare this problem with Example 10, whose additional complication makes a solution by slope-finding virtually mandatory.)

But for this problem, however, it would be far easier to use a computer (with "trial and error" solutions) if you have more towns to consider, not because any of the principles are different, but because the slope-finding gets too bogged down in ordinary algebra. For example, add to this problem only one more town, say town C, which is 6 miles south of the diner and 2 miles west. Call that distance to the water tower $d_3$. So the diagram now looks like this (Figure 17).

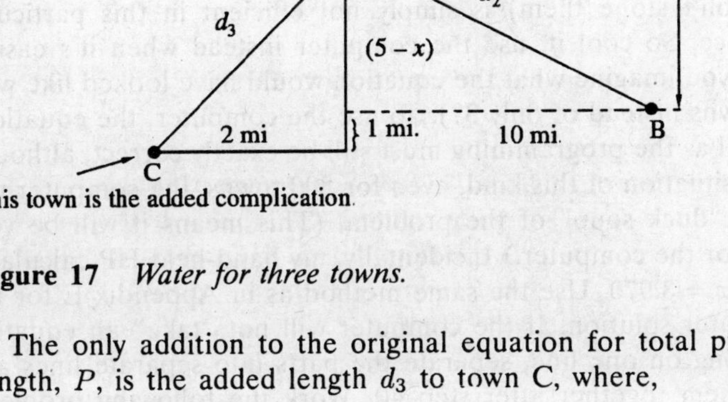


**Figure 17** *Water for three towns.*

The only addition to the original equation for total pipeline length, $P$, is the added length $d_3$ to town C, where,

$$d_3 = \sqrt{4 + (6 - x)^2}.$$

(Note that $5 - x + 1$ equals $6 - x$.) Now the full equation $P$, adding on $d_3$, is

$$P = (x^2 + 9)^{1/2} + (x^2 - 10x + 169)^{1/2} + (4 + (6 - x)^2)^{1/2}.$$

Slope-finding for this one is as easy as before;

$$\frac{dP}{dx} = \frac{1}{2}(x^2 + 9)^{-1/2}(2x) + \frac{1}{2}(x^2 - 10x + 169)^{-1/2}(2x - 10)$$
$$+ \frac{1}{2}(4 + (6 - x)^2)^{-1/2}(2x - 12).$$

Notice that $(6-x)^2$ is the same thing as $(x-6)^2$, since it becomes, when multiplied out, $x^2 - 12x + 36$. Its derivative is $2x - 12$.

Now when we let $dP/dx = 0$ and simplify slightly, we get

$$0 = \frac{x}{(x^2 + 9)^{1/2}} + \frac{(x - 5)}{(x^2 - 10x + 169)^{1/2}} + \frac{(x - 6)}{(x^2 - 12x + 40)^{1/2}}.$$

I am not embarassed to say that here I am stuck. The prodigious effort required to untangle the algebra is simply not worth it as long as you have access to a computer. You can program a problem like this (even with 20 towns), stuff it into a computer, and get your answer in a total time one hour at the most. So to use slope-finding (or, if some wish to use the word "calculus", we won't stone them) is simply not efficient in this particular instance. So cool it, use the computer instead when it's easier! (Can you imagine what the equation would have looked like with 20 towns instead of only 3?). To use the computer, the equations as well as the programming must still be exactly correct, although for a situation of this kind, even for 500 towns, the computer will make "duck soup" of the problem. (This means it will be very easy for the computer.) Incidentally, my hand-held HP calculator gives $x = 3.970$. Use the same method as in Appendix E for the computer solution. If the computer will not "take" an equation this long on one line, separate the parts into separate lines and add them together after step 40. Work the following problems out yourself, or even make up your own problem.

PROBLEM 7

There are three towns we may call points A, B and C, which make up an isosceles triangles (the sides are equal). Town A is north. Distances AB and AC are the sides, and are equal in length. Now the distance from town A, to the base BC is 20 miles, while the distance between towns B and C is 30 miles. Find the distance of the point P, south of town A, such that the sum of the distances to all the 3 towns A, B and C, to the point P shall be a minimum, and find this minimum distance of that point to the three towns. (The solution to this is in the Appendix A.)

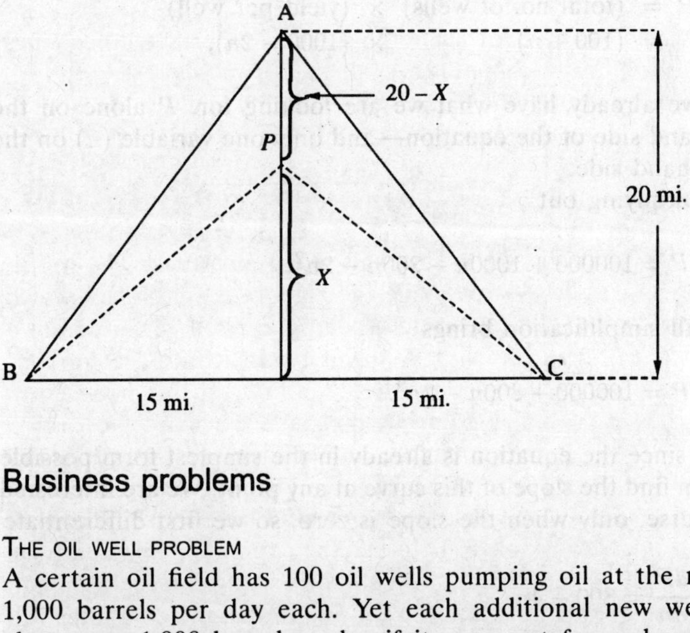


## Business problems

THE OIL WELL PROBLEM

A certain oil field has 100 oil wells pumping oil at the rate of 1,000 barrels per day each. Yet each additional new well will also pump 1,000 barrels a day if it were not for a drop in oil pressure in the entire field bringing a loss of 2 barrels a day for every well. If you wanted to extract the maximum yield from the entire field, how many new wells should you drill? Maybe even shut down some wells? What is the maximum production of the entire field?

Again, in order to find the slope of a curve, you must have the curve of the thing you wish to maximize, which in this case consists of the production of oil in the field for any number of wells. Now the production $P$, of the entire field is the product of the number of wells and the production per well (we are assuming that each well produces the same amount of oil, although a solution can easily be found for any combination of productions). So then, $P$ = (the total number of wells) × (yield per well). If we let $n$ = the number of new wells, then $100 + n$ must be the new total number of wells. And if each well was producing 1,000 barrels per day, but with each new well there is a drop of 2 barrels per day, then $1000 - 2n$ must be the new yield per well. So then

$$\begin{aligned} P &= (\text{total no. of wells}) \times (\text{yield per well}) \\ &= (100 + n) \qquad\qquad \times (1000 - 2n), \end{aligned}$$

and we already have what we are looking for, $P$ alone on the left-hand side of the equation—and only one variable ($n$) on the right-hand side.

Multiplying out,

$$P = 100000 + 1000n - 200n - 2n^2;$$

a small simplification brings

$$P = 100000 + 800n - 2n^2.$$

Now, since the equation is already in the simplest form possible, we can find the slope of this curve at any point. We are interested, of course, only when the slope is zero, so we first differentiate

$$\frac{dP}{dn} = 800 - 4n.$$

We let $\frac{dP}{dn} = 0$. Then,

$$4n = 800 \text{ and } n = 200.$$

Thus the number of new wells should be 200, for a total of 300 wells working in that field. To find the maximum number of barrels the field can yield per day, merely substitute our new found value of $n$ in the equation

$$P = (100 + n)(1000 - 2n).$$

Therefore, barrels per day (total production) is

$$P = 300 \times 600 = 180000.$$

Of course, it stands to reason that if you drill more wells than 200, you will get a situation called a "diminishing rate of return". If you graph $P$ versus $n$, this will become clear. Also, notice that to be ready to find the slope, the equation must be of the form with only one unknown ($n$) on the right-hand side. Now you do this problem.

PROBLEM 8

Suppose there is a fishing fleet in Norway going after herring. When there are fifty boats out, each boat brings in an average of 5,000 pounds of herring. But for each new boat over fifty, the fleet will experience an average loss of fifty pounds of herring per boat. Find the number of boats that should go out to catch the most herring for the fleet, and find this maximum catch. Before you do this problem you might ask, "but is this a real problem? Aren't there more things involved than just the maximum yield of the fleet? What about the cost of the new boats, the cost of the gasoline, and the cost of the men's wages, extra gear, added insurance?" The answer is this: the principle is the same, no matter how many factors are involved. Just as long as you can write them in mathematical notation and relationships, you could do 100 factors like cost of gasoline, wear and tear on the engines etc. etc., and find the maximum profit if you like. (The answers can be found in Appendix A.) To illustrate this, after you do this problem, add to it the cost of additional manpower. Let us say that for each boat over fifty, the cost in additional manpower in its "herring-take" equivalent is 500 pounds, and the gasoline expense for each additional boat, in terms of the herring, is 50 pounds. We will have to specify the problem in terms of one variable, anyway. So we can let all the added cost factors be related to the catch as above. This is the principle, no matter how many details have to be taken into account. We do recognise, however, that sometimes adding new factors makes the problem so cumbersome that it will be virtually unsolvable without the computer such as the "water for 3 towns". As we leave this section on slope-finding, let us review some of the principles that are involved.

## Summation of slope-finding

1. Calculus is really *slope-finding*, and you should not let the word "calculus" frighten you away from this. Once you understand that we are finding the general slope, the actual math does not require complex principles, but only attention to details.

2. Why do we need the $\Delta x$ to use with the $x$ to find the slope at a point? Can't we do it with just the $x$?

   The reason we need $\Delta x$ is because one point ($x$) is used to sight mark the horizontal line of the slope triangle, the other point ($\Delta x$) is used to sight mark the vertical line of the slope triangle. You cannot measure a slope with only one point (as one line), it must be two points for two lines. We let $\Delta x$ be the smallest number you can think of (without being zero) that is nearest $x$. Therefore, $\Delta x$ is next to $x$ no matter how sharp the curve. Thus we must use $x$ and $\Delta x$ to establish the slope of a curve at a point on it.

3. Whatever the problem, whether it is maximizing the profit or minimizing area, or whatever, the description must be put into equation form. Obviously, we can't find the first derivative of an equation without the equation to start with.

4. The equation should have the thing you wish to find, maximum profit, etc., alone on the left-hand side of the equation, while on the right-hand side must be the description of the problem in terms of only one unknown. Usually, geometric, etc., relationships will allow substitutions until this is so.

5. For graphic representation and visualization, the item alone on the left-hand side of the equation is the ordinate (the up and down axis—the $y$-axis), while the item on the right-hand side is the abscissa (the horizontal axis—the $x$-axis). This is how it should be visualized, expecially if you don't want to graph the problem.

6. Many problems using slope-finding for finding maximum or minimum values can almost always be also done without the use of slope-finding, (that is without using calculus), although, usually more effort would be involved if one were to do the problem by "trial and error" or by accurate graphing. The computer can be very useful in many cases, but is useless in others.

7. When we find the slope, or differentiate, as calculus texts call it, we are automatically finding the slope for all the points of the equation (just like an equation automatically gives all points on its own curve). If we were looking for maximum or minimum values, then that is where the slope is zero. We have to let the derivative (the slope) be equal to zero. The location (on the horizontal axis) will be found by this means.

8. The hardest part of slope-finding, or calculus, as it is called, is setting up the equation correctly and in the desired form. Slope-finding is just about the easiest part of it.

9. Slope-finding (or differentiation) is not only a particular operation that you can do to an equation, but it is also a true mathematical operation in its own right, just like multiplication or addition. So you did know four basic operations: addition, subtraction, multiplication, and division. But if there is an opposite to addition (subtraction) and also an opposite to multiplication (division), and if differentiation is a true mathematical operation in its own right, there will be an opposite to it as well. So you now know five, coming up the sixth!

# 2

# Integral Calculus
## Area-Finding

The above title is only a fancy pair of words that means "finding an area". What is the need for finding areas? There are many surprising uses. For example, the area under a curve in economics may represent profit, or represent loss. In engineering, the area under a particular curve may represent the energy of a bullet or the work of a machine or the deflection of beams, the like of which are used in every structure, or the location of the center of gravity of a spacecraft. In physics, the area under a curve may be made to represent the energy yielded from an atomic blast, or from a tracer element in medicine. All the other sciences have found innumerable uses for it.

Yet, these are only a minute number of the uses for finding the area under a curve. New applications for area-finding (integration) are being found all the time. And one of the most useful and astonishing uses of this area-finding process is the fabulous way that the formulas for the Laws of Nature for every single field of endeavour can be found! The way that this is done is through the *Differential Equation!* Near the end of this book, there will be more on that.

### A method for area-finding

Now there are certain figures whose areas can be very easily found, particularly figures enclosed in straight lines. (We are only considering plane figures for the time being.)

1) The *rectangle* (the four-sided figure)—The area is the product of the base and the altitude.

2) The *triangle*—This area is one-half the product of the base and the altitude.

With the following two familiar figures, the boundaries are not straight lines, but the formulas for finding the areas are known.

1) The *circle*—The area is the product of Pi and the radius squared ($A = \pi r^2$).
2) The *ellipse*—Its area $A = \pi ab$, where $a$ and $b$ are the semi-major and semi-minor axis.

This is all well and good for these very used figures. But what do we do for finding the area involving other curves? For example—how can we find the area under the curve $y = x^2$ (Figure 18) say, between $x = 0$, and $x = 2$? We choose the function $y = x^2$ only for the ease of illustration, as before.

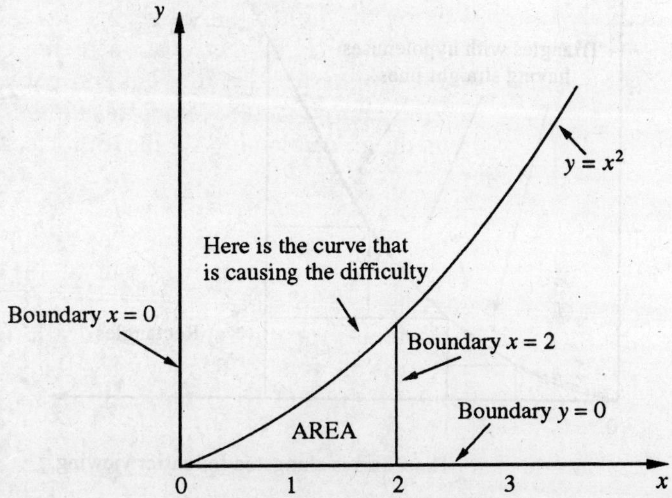


**Figure 18** *How can we find the area under the curve $y = x^2$, if the boundaries are not all straight lines, or parts of circles or ellipses?*

From here on I want to warn you that the following sequence may seem exceedingly simple, too simple, but watch out because it is simple except that several simple things will be happening, all at the same time.

Now, we can do certain things like:

(*a*) Divide the area in question into rectangles and triangles, whose areas we can measure (Figure 19). Then we could add them all up. This is not a bad idea for an approximation. But if we want to be more accurate, using more (smaller) rectangles and triangles, this can get very involved and tedious for each curve that we consider. Also, the area under the curve of $y = x^2$ (or any other) still does not fit a series of triangles perfectly (although the rectangles and triangles themselves can be computed exactly) since the hypotenuses of the little triangles should not have curved lines no matter how small the triangles be.

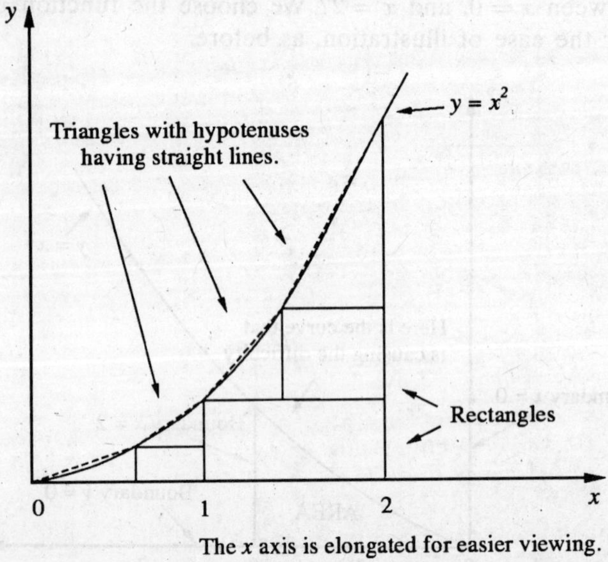


**Figure 19** *Graph of $y = x^2$. We are dividing up the area into several rectangles and triangles.*

(*b*) There are hand operated instruments that measure area. One is called a planimeter. There are also computer methods for determining areas. But let us see if we can develop a method that can be applied to all cases, and one that is better than using instruments.

Firstly, let us try to approximate the area under the curve of $y = x^2$, between $x = 0$ and $x = 2$, with one rectangle (Figure 20) by judging the height with our eyes. (The trouble is that no two people will judge exactly identical heights.)

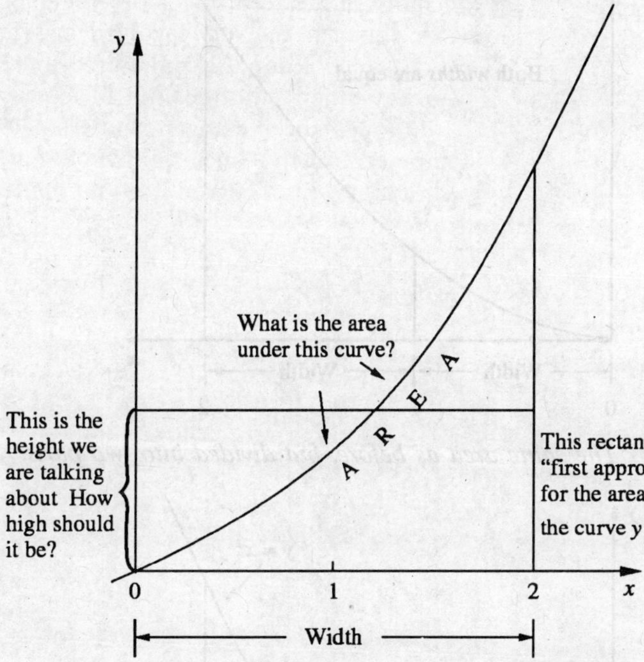


**Figure 20** *Approximating the area under the graph of the curve $y = x^2$ between $x = 0$ and $x = 2$ with one rectangle.*

While the width 2 is very simple to see, the height to make the rectangle equal in area is something else. Nobody has ever found out how to judge the height of the rectangle such that its resulting area will be identically equal to the area under the curve. So we have a problem which appears unsolvable at first glance. But let us, then, divide the area under the curve into two pieces of equal width (Figure 21).

Now if we use two rectangles (Figure 22), one for each piece, to approximate each area, this method appears to make the job somewhat easier because we cannot *judge* the heights *worse* than the case before when we only had one width. That is, we have now divided one large problem into two smaller ones.

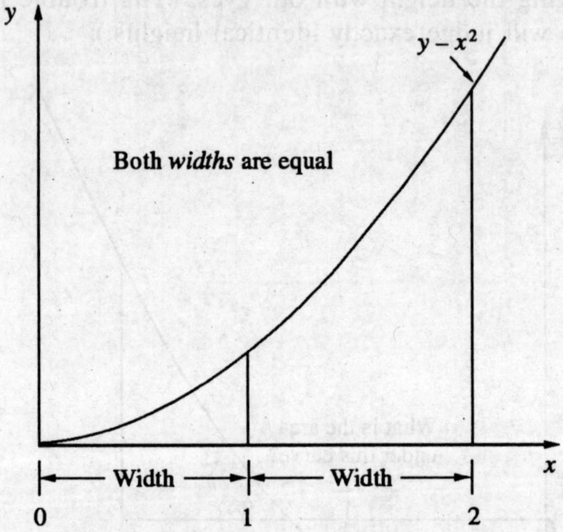


**Figure 21** *The same area as before, but divided into two parts.*

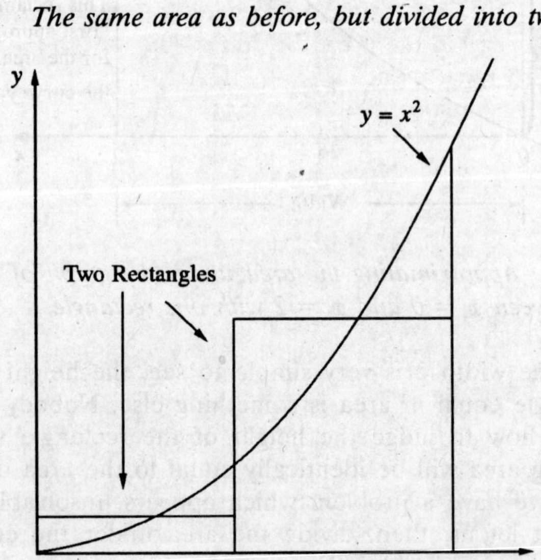


**Figure 22** *Both rectangles approximate the area under the curve, their heights are something like "average" heights of the curve.*

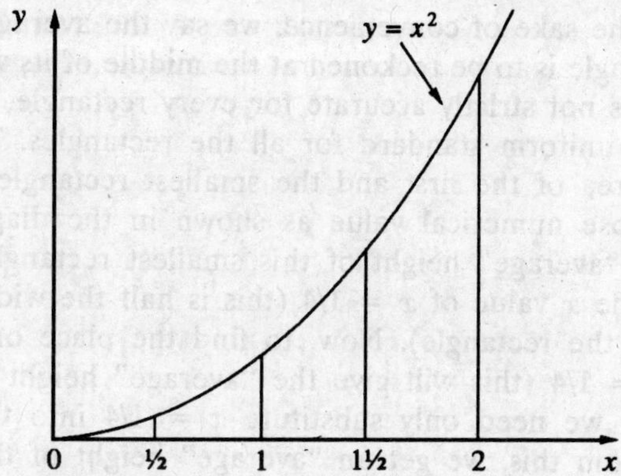


**Figure 23** *The same area, but this time divided into four pieces, each of equal width.*

So much for two pieces. Let us try four pieces, since it seems that we might be getting somewhere (Figure 23).

This time, for convenience, let us take the "average" height of each rectangle to be at the *mid-point* of each of the four widths (Figure 24), although you can imagine that each rectangle is *only* a rough average of the area! But the object is still the same—to make the rectangles such that their areas are equal to each of the four pieces under the curve of $y = x^2$. And we have to have something better than judging by eye, since we need something uniform—a method that is not different for each person!

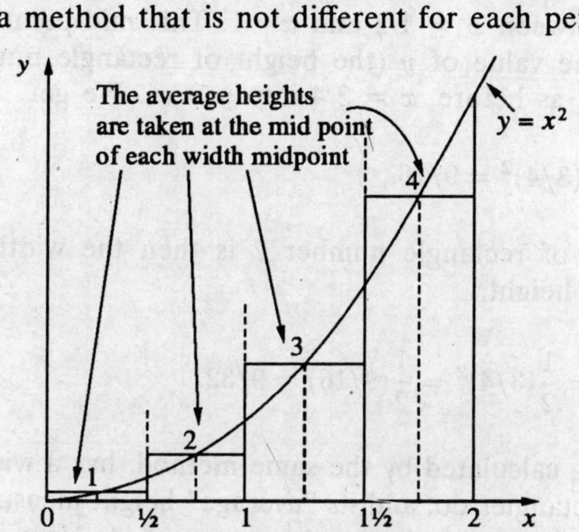


**Figure 24** *Four rectangles, each width is the same.*

If, for the sake of convenience, we say the average height of each rectangle is to be reckoned at the middle of its width, while we see it is not strictly accurate for every rectangle, at least we can use a uniform standard for all the rectangles. This means that the area of the first and the smallest rectangle will be its width, whose numerical value as shown in the diagram is 1/2 times the "average" height of this smallest rectangle, which is taken at the $x$ value of $x = 1/4$ (this is half the width, and the middle of the rectangle). Now, to find the place on the curve where $x = 1/4$ (this will give the "average" height of the first rectangle), we need only substitute $x = 1/4$ into the function $y = x^2$ (from this, we get an "average" height of the first and smallest piece of the curve $y = (1/4)^2 = 1/16$). So the area 1 (representing rectangle 1) = (width of area 1) × (average height of area 1)

$$A_1 = \frac{1}{2}(1/4)^2 = \frac{1}{2}(1/16) = 1/32.$$

Now we'll calculate area 2 in the same way. (We are using a uniform method of estimating the average heights of each rectangle.) The width is still 1/2, because we are using only identical widths. The "average" height for area 2 is to be calculated only at the mid-point of rectangle number 2, whose width occurs exactly between $x = 1/2$ and $x = 1$. This mid-point is $x = 3/4$. To find the value of $y$ (the height of rectangle number 2), we substitute, as before, $x = 3/4$ into $y = x^2$. We get

$$y = (3/4)^2 = 9/16.$$

The area of rectangle number 2 is then the width of area 2, times the height,

$$A_2 = \frac{1}{2}(3/4)^2 = \frac{1}{2}(9/16) = 9/32.$$

Area 3, calculated by the same method, has a width of 1/2 as all the rectangles do, and its "average" height measured between $x = 1$ and $x = 1\frac{1}{2}$, is $x = 1\frac{1}{4}$ or $x = 5/4$. Now the height is measured, as before, by substituting $x = 1\frac{1}{4}$ into the equation $y = x^2$, obtaining $y = (5/4)^2 = 25/16$. The third area, $A_3$, is then

$$\begin{aligned} A_3 &= \text{width} \times \text{average height}_3 \\ &= \frac{1}{2}(5/4)^2 = \frac{1}{2}(25/16) = 25/32. \end{aligned}$$

Now, for area 4, similarly, we obtain the average height (which is when $x$ is between $x = 1\frac{1}{2}$ and $x = 2$). The height then is $y = (7/4)^2 = 49/16$. This area is then

$$\begin{aligned} A_4 &= \text{width} \times \text{average height}_4 \\ &= \frac{1}{2} \times (49/16) = 49/32. \end{aligned}$$

The total area ($A_T$) is: $A_T = A_1 + A_2 + A_3 + A_4$ (for the four pieces).

Now substituting the actual numbers,

$$A_T = \frac{1}{2}(1/16) + \frac{1}{2}(9/16) + \frac{1}{2}(25/16) + \frac{1}{2}(49/16).$$

This is from

$$A_T = \underbrace{\frac{1}{2}(1/4)^2}_{\texttt{1st term}} + \underbrace{\frac{1}{2}(3/4)^2}_{\texttt{2nd term}} + \underbrace{\frac{1}{2}(5/4)^2}_{\texttt{3rd term}} + \underbrace{\frac{1}{2}(7/4)^2}_{\texttt{4th term}}.$$

(Notice that the 7 in the numerator is one less than twice the number of terms.)

Notice also that the 4 in the denominator is common to all of the terms. This is a big advantage that we enjoy using a uniform method for estimating the height. Also, note that the progression in the numerator is $1^2, 3^2, 5^2, 7^2$. I think we may be able to do something with this also.

What do you think?

The total area ($A_T$), calculated in this fashion by rough estimation of heights, turns out to be

$$A_T = \frac{1}{32}(1^2 + 3^2 + 5^2 + 7^2) = \frac{1}{32} + \frac{9}{32} + \frac{25}{32} + \frac{49}{32},$$

and the area calculated from four rectangles,

$$A_T = 84/32 = 2.625.$$

Now let us divide the sought-after area into 8 equal widths (Figure 25). I think you are starting to see something that is going to happen. First of all, the method we are using to estimate the height of each rectangle is starting to look better and better as we go along, that is, we are calculating the true area more and more accurately.

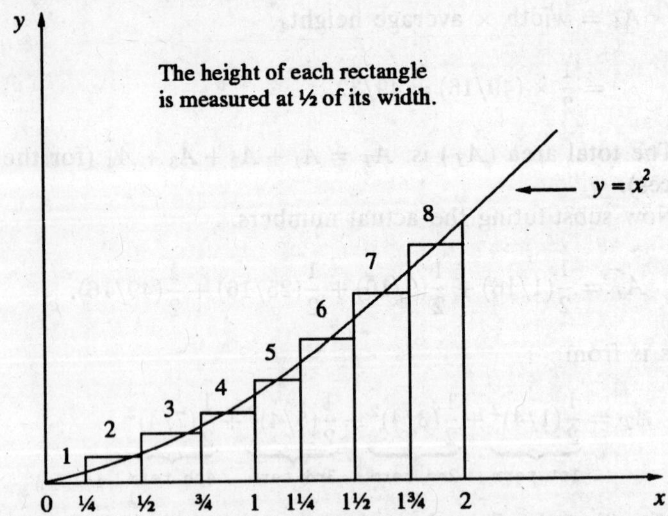


**Figure 25** *The total area is divided into eight rectangles, each having equal width (but not equal height, of course).*

Can you see that the average heights are getting more and more accurate as the slices are getting thinner as we proceed? That's a big help. And it gets more interesting as we proceed.

Now, since we are using 8 pieces, the widths now are each 1/4, obtained by dividing the total length 2 by the number of pieces, i.e., 8, which is 2/8 or 1/4 for each width. (We are going to use this idea again later on. It is an important detail.)

Now to calculate each height! We take the first width, 1/4, and divide it in half. We are now always going to fix permanently the average height at the middle of each width, or in this case $1/4 \times 1/2 = 1/8$. To find the average height of area 1, we substitute $x = 1/8$ into $y = x^2$, or $y = (1/8)^2$.

Then $A_1$ = (width) × (average height of rectangle 1)

$$A_1 = (1/4) \times (1/8)^2.$$

For $A_2$ = (width) × (average height of rectangle 2)

$$A_2 = (1/4) \times (3/8)^2.$$

For $A_3$ = (width) × (average height of rectangle 3)

$$A_3 = (1/4) \times (5/8)^2.$$

Notice that the numerator of the average heights starts with 1 and increases by 2 each time so that it is easy to see that this sequence will follow:

$$A_4 = (1/4) \times (7/8)^2,$$

$$A_5 = (1/4) \times (9/8)^2,$$

$$A_6 = (1/4) \times (11/8)^2,$$

$$A_7 = (1/4) \times (13/8)^2,$$

and $A_8 = (1/4) \times (15/8)^2.$

Putting them all together, we have

$$A_T = (1/4)(1/8)^2 + (1/4)(3/8)^2 + (1/4)(5/8)^2 + (1/4)(7/8)^2$$
$$+(1/4)(9/8)^2+(1/4)(11/8)^2+(1/4)(13/8)^2+(1/4)(15/8)^2.$$

Or, we could have it this way as well:

$$A_T = (1/4) \times (1^2/64) + (1/4) \times (3^2/64) + (1/4) \times (5^2/64)$$
$$+ (1/4) \times (7^2/64) + (1/4) \times (9^2/64) + (1/4) \times (11^2/64)$$
$$+ (1/4) \times (13^2/64) + (1/4) \times (15^2/64).$$

Now we can, and should, factor out the 1/4 and 1/64, since it is common to each term,

$$A_T = (1/4) \times (1/64)(1^2 + 3^2 + 5^2 + 7^2 + 9^2 + 11^2 + 13^2 + 15^2).$$

Notice that 15 is one less than twice the number of terms. The total area, using the eight pieces:

$$A_T = (1/256)(1 + 9 + 25 + 49 + 81 + 121 + 169 + 225)$$
$$= 680/256 = 2.656.$$

(Compare this with the previous area of 2.625 where we used 4 pieces.) Another interesting thing is that as we increase the number of pieces, the resulting computed total area increases (as from 4 pieces we obtained $A_T = 2.625$, while from 8 pieces we obtained $A_T = 2.656$). Now we know that if we increase the number of pieces, the computed total area will be increased. But we also know that the true area has to be only one true answer, and that area cannot keep increasing indefinitely! After all, the number of pieces may be increasing, but the area into which they are being fitted, cannot be increasing nor changed in anyway just because the fit of these pieces is getting better and better. So what we are doing is chasing the largest number of pieces possible, that will give the true area, the area which will not increase any longer. And what could be the largest number? It could only be infinity, $\infty$ (this is the notation for infinity). But how can anyone do an infinite number of operations? Even 1,000 pieces would fatigue anyone. The computer is immune from fatigue. But we are looking for a general formula. The computer is useless here since it can only give numbers. We will use the computer to check the results, however. So follow this:

Now we know that if we slice up the original area under the curve of $y = x^2$ into, say, 16 pieces, we will end up with the same type of formula again. There will be 16 terms, the first having $1^2$ in the numerator, the second $3^2$, and the third $5^2$ and so on for 16 terms. And the more slices, the more *accurate* is going to be the representation. I don't think anyone would quarrel with that assessment! We would like to, if possible, slice the area into 1,000 pieces and even more if it wasn't such a nuisance adding up all the terms.

## Here comes another device (courtesy of Sir Isaac Newton)

So, in place of 1,000 slices, let us slice up the area into $n$ slices. In place of $x = 2$, let us have an overall length of $x = b$ (Figure 26).

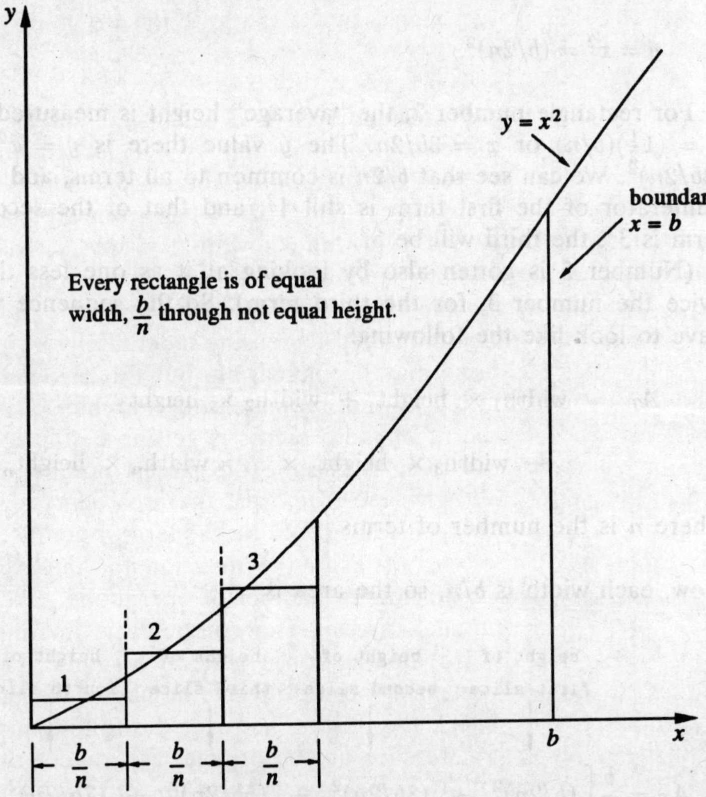


**Figure 26** *A device for finding the area under the curve $y = x^2$. We are replacing the dimension numbers with letters. The number of n slices is not further specified for now. That is, why we show only a few slices, each of width $b/n$ is because the number of slices is far too many for us to show them individually. But every slice has the same width $b/n$.*

Now, each individual width is the full length $b$, divided by the number of slices, $n$, which is $b/n$. There is no other change at all; everything is calculated as was done before. Halfway along the first rectangle is 1/2 of $b/n$, which is $b/2n$.

The average height of rectangle number 1 is obtained the same way as before, that is, by substituting $x = b/2n$ (halfway along the first width) into $y = x^2$. We get, by this means,

$$y = x^2 = (b/2n)^2.$$

For rectangle number 2, the "average" height is measured at $x = (1\frac{1}{2})(b/n)$ or $x = 3b/2n$. The $y$ value there is $y = x^2 = (3b/2n)^2$. We can see that $b/2n$ is common to all terms, and the numerator of the first term is still $1^2$, and that of the second term is $3^2$, the third will be $5^2$.

(Number 5 is gotten also by looking at it as one less than twice the number 3, for the third term). So the sequence will have to look like the following:

$$A_T = \text{width}_1 \times \text{height}_1 + \text{width}_2 \times \text{height}_2 + \text{width}_3 \times \text{height}_3 \times \ldots \times \text{width}_n \times \text{height}_n,$$

where $n$ is the number of terms.

Now, each width is $b/n$, so the area is

height of first slice → $(b/2n)^2$; height of second slice → $(3b/2n)^2$; height of third slice → $(5b/2n)^2$; height of fourth slice → $(7b/2n)^2$

$$A_T = \frac{b}{n}\left[(b/2n)^2 + (3b/2n)^2 + (5b/2n)^2 + (7b/2n)^2 + \ldots + [(2n-1)b/2n]^2\right].$$

↑ This is one less than twice the number of terms $n$, the height of the last slice

(The dots mean that the terms continue in the same form to the last term, which is shown.)

The coefficients of the numerators are always one less than twice the number of its term. For example, in term 2, the 3 is gotten by subtracting 1 from twice 2 or $(2 \times 2) - 1 = 3$. Each numerator is thus obtained. Now you can see that we ought to factor out of each term that is common to all, namely $(b/2n)^2$. Doing this we get

$$A_T = (b/n)(b/2n)^2[1^2 + 3^2 + 5^2 + 7^2 + \ldots (2n-1)^2].$$

We just multiply together the outside coefficients:

$$A_T = b^3/4n^3[1^2 + 3^2 + 5^2 + 7^2 + \ldots (2n-1)^2].$$

Very fortunately, there is a very convenient formula which is exactly equal to the *sum* of the sequence in the bracket, namely

$$1^2 + 3^2 + 5^2 + 7^2 + \ldots (2n-1)^2 = \frac{4n^3}{3} - \frac{n}{3}. \quad \texttt{This is it!}$$

You can easily check it. For example, if we take only the first 4 terms, and add them up,

$$1^2 + 3^2 + 5^2 + 7^2 = 1 + 9 + 25 + 49 = 84. \quad \texttt{Answer}$$

Using the short cut formula $(4n^3/3) - (n/3)$, we substitute $n = 4$ for the number of terms we are considering. We get

$$4(4)^3/3 - 4/3 = (4/3) \times 64 - (4/3) = 4/3(64 - 1)$$

$$= (4/3) \times 63 = 84. \quad \texttt{Exactly the same answer}$$

So we can use this very convenient formula, $[(4n^3)/3 - n/3]$, as an exact replacement of that number sequence, which we do:

$$A_T = (b^3/4n^3)[1^2 + 3^2 + 5^2 + 7^2 + \ldots (2n-1)^2]$$

$$= (b^3/4n^3)[4n^3/3 - n/3].$$

Now multiply through by only the denominator $4n^3$. We get

$$A_T = b^3\left[\frac{1}{3} - \frac{1}{12n^2}\right] \quad \text{for the final step.}$$

If we want to use 1,000 terms, equivalent to 1,000 pieces, that is $n = 1{,}000$, then it is very clear that the magnitude of $1/12n^2$ will be very, very small in comparison to the value 1/3, because 1,000 is in the denominator *and is squared,* making the value of that term virtually nothing! And if we wanted to use $n = 1{,}000{,}000$ slices instead of 1,000, then the magnitude of $1/12n^2$ in comparison to 1/3 will be even more incalculably small. So, I think you see that we lose literally nothing by eliminating the $1/12n^2$ term. In fact we gain in accuracy because more the slices used, the smaller in value that term becomes. Here is where we use an "infinite" number of pieces at last! This, then, is the beauty of that particular step! We then have a total area $A_T = b^3/3$.

Now, instead of using a particular length $b$, what if we use a general length, meaning any length $x$ ? Then we should replace $b$ with $x$, and the total area will become

$$A_T = \frac{x^3}{3}. \quad \longleftarrow \quad \text{This is the general formula for the area under the curve } y = x^2$$

In the first part of this book we have found how to transform the expression. $y = x^2$ such that we can find the *slope* at any part of this curve by another expression $dy/dx = 2x$. Now we have just found out how we can transform the expression $y = x^2$ such that we can find *the area* of any portion of that curve with another expression, $A = x^3/3$.

So this is what it's all about, what all the fuss about calculus is—finding the expressions for slope, and for areas. But in having the ability to find these expressions, something new and unexpected will happen. It will be just as startling and unexpected as a caterpillar that turned into a butterfly. But that is ahead of the story. So let us continue...

Now, if we wanted to calculate the area up to the line $x = 2$, we substitute 2 for the $x$,

$$A_T = 2^3/3 = 8/3 = 2.666\ldots$$

This is exactly the correct answer for the area under the curve $y = x^2$ between $x = 0$ and $x = 2$, and cannot be improved upon no matter how many slices are used. Do you agree? This answer compares with these other values of $n$,

| Number of slices | Area | |
|---|---|---|
| 4 | 2.625 | |
| 6 | 2.648 | |
| 8 | 2.656 | |
| 20 | 2.665 | |
| $n$ | 2.666. . . | This area is exactly correct = $2\frac{2}{3}$ |

↑

Letting $n$ be a very large number or even better, infinity

Notice how the accuracy improves with more slices.

Now, for your own pleasure and for your own proof, you should graph very accurately the function $y = x^2$, say between $x = 0$ and $x = 2$, on quite a large graph paper to check this result.

Graph and plot $x$ versus $y$ at $x = 0.1$, 0.2, 0.3, etc. Make sure the $y$ and $x$ values and the curve are very accurately placed. Then add up the boxes on the graph paper by closely estimating the heights of each and every skinny rectangle. Do 20 slices, or even double that if you have the patience. See how close to the area of 2.66666. . . you can actually get. I think you will be amazed that this method works so very well. In fact, this answer is so perfect that no amount of slices can improve upon it. Try it.

So, for the function $y = x^2$, we have the area, $A = x^3/3$, which is the general formula of area for the function. But what about other functions, such as $y = 3x^3 - 2x^2 + x - 1$, for example? Can this method be applied here, too? Will the sequence $1^2, 3^2, 5^2, 7^2 \ldots$ be different? Will another unforeseen complication arise to confound us? Will the method then fail?

The answer is, that the identical procedure does not run into unforeseen complications, and it encounters no difficulties! Yes, of course this method works. But instead of having to go through the very long-winded method that we have just done again, we need only remember a simple rule (or a short cut device) for finding the area of $y = x^n$. It is: Raise the exponent of $x$ by one, and divide the term by that resulting exponent (i.e., place the resulting number in the denominator), so that the general rule for this type of formula is:

If we start with the type $y = x^n$,

where $n$ is any number (except $-1$). If $n = -1$, and we raise the exponent by 1, i.e., we add 1, we would be dividing by zero. This we cannot do.

$$\text{Then, Area} = \frac{x^{n+1}}{n+1}.$$

(The boundaries of the curves will also be dealt with soon.)

Now, if we have several terms in the equation, for example

$$y = 3x^3 - 2x^2 + x - 1,$$

we treat each term individually and raise the exponent of $x$ by 1, then put the resulting exponent into the denominator. For $y = 3x^3 - 2x^2 + x - 1$, this now becomes,

$$\text{Area} = \frac{3x^4}{4} - \frac{2x^3}{3} + \frac{x^2}{2} - x + C.$$

Notice that the 1 becomes $x$ because 1 is actually $x^0$, and adding one to zero makes $x^1$ ($x$ to the first power, or just plain $x$). ($C$ is a constant which accompanies all integrations that do not have definite boundaries specified.)

A very brief table of integrals (short cut formulas for different forms of equations) is given in Table 2.1 later.

## The relationship between slope-finding and area-finding

This process of finding areas is exactly the reverse of finding the slope. For finding the slope, you will remember, we *multiplied* the coefficient by the exponent, and then subtracted the exponent by 1. For instance, using the previous area $y = x^3/3$, if we find the slope (differentiate), we get $dy/dx = x^2$, which is the function we started with. For area-finding, we raise the exponent by one and divide by that resulting number. Why area should be related mathematically to slope at all is left to your imagination.

## The notation used for area-finding

Now, there is an agreed-upon notation that is used for the process of area-finding. The first is the symbol $\int$ which stands for "sum" (from the old English and German S), which in turn means to us now, "the sum of an unspecified very large number of $n$ rectangles under the curve," and $dx$ stands for the very small width of each rectangle. So that for the area between $x = 0$ and $x = 2$ of the function $y = x^2$, we write

$$A \doteq \int_0^2 x^2 dx.$$

The zero at the bottom of the integral sign stands for the value of the boundary on the *left* side of the curve (these two numbers placed up and down the integral sign are also called the limits), and the 2 on the top of the integral sign stands for the boundary on the right side.

If this diagram could speak (Figure 27), it would say that the area under the curve $y = x^2$ from the boundaries $x = 0$ to $x = 2$ is the sum of the areas of $n$ number of rectangles of height $y$, and very thin width $dx$ (so the *area* of each thin rectangle is $ydx$), when the number $n$ is very large (say, at least 1,000).

$A = \int_0^2 ydx.$ This says the same thing using mathematical symbols

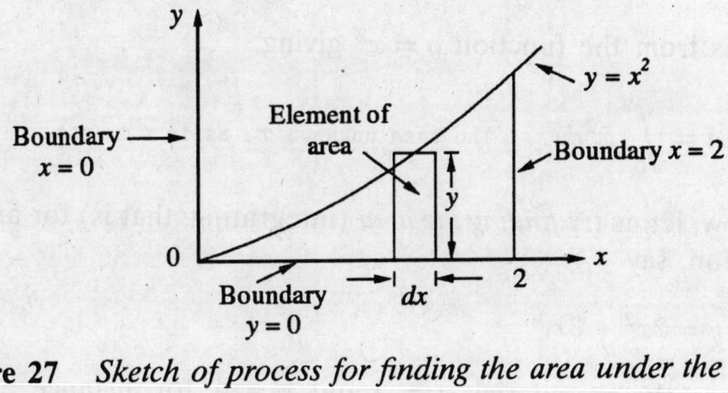


**Figure 27** *Sketch of process for finding the area under the curve.*

Here $y$ is the height, $dx$ is the width of a typical element of area under the curve. We sum up, by this short cut process, all the elements of areas (the larger the number, the better). We substitute $x^2$ for the $y$ from $y = x^2$, between the symbol and $dx$.

$$A = \int_0^2 x^2 dx.$$ ← The $dx$ refers to a very, very large number ($n$) of very thin slices ($n = \infty$)

To denote that we have integrated the term $x^2$ and should evaluate it between the limits $x = 0$ and $x = 2$, we write

$$A = \left[\frac{x^3}{3}\right]_0^2,$$

which we calculate in this way:

$$\frac{(2)^3}{3} - \frac{(0)^3}{3} = \frac{8}{3}.$$ Area! We have replaced $x$ first with the upper limit (2), then the lower limit (0)

We substitute the 2 where $x$ is, then the zero, after which we subtract the latter term (or limits) from the former. The answer then is 8/3. The process is called integration, or finding the area, or anti-differentiation. Notice that we do not integrate until we substitute for the $y$ in

$$A = \int_0^2 y dx.$$

This is from the function $y = x^2$ giving

$$A = \int_0^2 x^2 dx.$$ The same unknown $x$, as in $dx$, etc.

Now, let us try *finding the area* (integrating, that is) for another function, say

$$y = 2x^2 - 3x$$

between the boundaries $x = 2$ and $x = 4$, for instance.

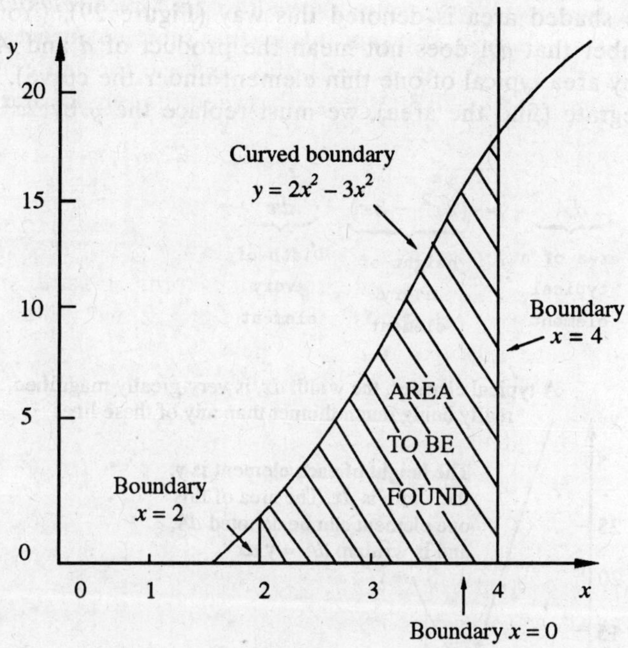


**Figure 28** *Finding the area under the curve* $y = 2x^2 - 3x$ *(shaded area).*

We should sketch this function within these boundaries for visualisation. We make a table between $x = 2$ and $x = 4$.

Table for finding several points of the function $y = 2x^2 - 3x$.

| $x$ | $2x^2 - 3x$ | $y$ |
|---|---|---|
| 2 | 8 − 6 | 2 |
| 3 | 18 − 9 | 9 |
| 4 | 32 − 12 | 20 |
| $2\frac{1}{2}$ | 12.5 − 7.5 | 5 |
| $3\frac{1}{2}$ | 24.5 − 10.5 | 14 |

Please watch your arithmetic very, very carefully, since, by far the hardest thing about calculus is preventing mistakes in arithmetic!

The curve of the function $y = 2x^2 - 3x$, then looks like this (Figure 28).

The shaded area is denoted this way (Figure 29). (You will remember that $dA$ does not mean the product of $d$ and $A$, but is a tiny area typical of one thin element under the curve). Now, to integrate (find the area), we must replace the $y$ by $2x^2 - 3x$ giving

$$\underbrace{dA}_{\text{area of a typical element}} = \underbrace{(2x^2 - 3x)}_{\text{height of every element}} \underbrace{dx}_{\text{width of every element}}.$$

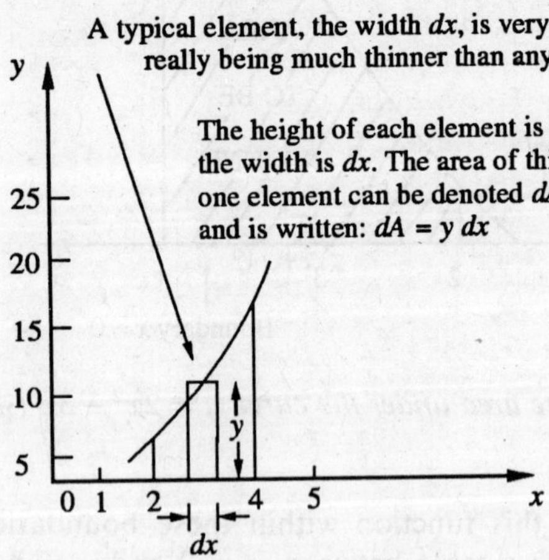


**Figure 29** *ydx, An element of area.*

Now, summing up all thin areas ($dA$ between $x = 2$ and $x = 4$) we further write it as

$$\int_2^4 dA = \int_2^4 (2x^2 - 3x)dx.$$

We denote the sum of the $dA$'s by just plain $A$

$$A = \int_2^4 (2x^2 - 3x)dx.$$

We continue by integrating term by term (of course there are only 2 terms here),

$$A=\left[\frac{2x^3}{3}-\frac{3x^2}{2}\right]_2^4.$$

`This completes the integration, but has not yet dealt with the boundaries, which will then give a numerical value`

To check that you did the integration correctly, simply reverse the process and differentiate the above terms, and you must get the expression you started with, that is, $2x^2-3x$, because differentiation is the *exact opposite* of integration. Now to evaluate this expression for area, first substitute the upper limit, which is 4, for the $x$ (this stands for the boundary on the right, by international convention), then subtract the same term, but with the lower limit substituted, which is 2 (this stands for the boundary on the left), like this:

$$A \;=\; \left[\frac{2(4)^3}{3}-\frac{3(4)^2}{2}\right] \;-\; \left[\frac{2(2)^3}{3}-\frac{3(2)^2}{2}\right].$$

`Area up to the boundary on the right side from the origin. Here we have substituted 4 for x`    `Area up to the boundary of the left side from the origin. Here we have substituted 2 for x`

We get, then,

$$A=\left(\frac{128}{3}-\frac{48}{2}\right)-\left(\frac{16}{3}-\frac{12}{2}\right).$$

(If the terms in the brackets are many, it is very important not to drop the brackets or deal with the negative sign between the brackets until everything in the brackets is clarified, for avoiding arithmetic errors.)

We see that it is worthwhile to drop the brackets at this point, because the combining of terms is simplified;

$$\text{Area, } A=\frac{128}{3}-\frac{16}{3}-\frac{48}{2}+\frac{12}{2}=\frac{112}{3}-\frac{36}{2}=19\frac{1}{3}. \qquad \textbf{Answer}$$

For the process of finding areas under curves (integration), there are sets of "cook book" formulae, but unfortunately, one needs a separate formula for every case that is the slightest bit different in form from the other. That is why there are thousands of formulas for integration that can be looked up, and there are many, many thousands more for which, unfortunately, no known "cook book" formulas exist. I say unfortunately, because there are hordes of complicated formulas that are very important in Engineering and all branches of Science that must be tackled by other means that are not at all satisfactory. Even the most modern high-speed computers are over-taxed by these problems. For example, suppose we want to integrate an equation for which no known solution exists. Then we must resort to a numerical method (Simpson's rule, etc.) which is an "adding-up" of many thin rectangles that would make up the area. Very frequently, the significance of the answer to a problem completely depends on how accurately the area can be calculated. Suppose then, we use 1,000 rectangles to find a certain area. That would be very accurate! This means that the integrand (the term to be integrated) has to be calculated 1,000 times, each time with a slightly different value of $x$. Well, the computer is very fast and it can arrive at an answer, say, in a few seconds. But we are working only in 2 dimensions thus far, $x$ and $y$. Most real problems in Engineering and Science are in 3 dimensions (double integrals ($x$, $y$, $z$)) or in 4 dimensions (triple integrals ($x$, $y$, $z$ and $t$, for time)) or even more. If we are using 1,000 divisions for the area, then for each of the thin rectangles in the $x$, $y$ direction, we need to do the problem 1,000 times (for the $z$ direction), i.e., each new dimension multiplies the problem's computations by 1,000 times. And what if that equation is only one in a series of equations that make up the whole problem? I think you see what is involved now! But had these formulas been integrable in "closed form" (i.e. a solution is known), then the problem could be done in seconds. But mostly, we really need a function, i.e., an equation, not just a series of numbers! That is why integration (anti-differentiation) is the greatest problem in calculus, rather than slope-finding (differentiation). It seems that an equation can almost always be managed to be differentiated, but not so, integrated. Needless to say, when we cannot solve a certain integral in closed form, we cannot obtain the general formula for its area in simple form. We can only get a numerical answer for that particular integral. For many cases this is good enough. Anyway, there's not much else that can be done about it, except for finding more "closed form" solutions. Table 2.1 lists some of the typical forms. Any math or science handbook will have hundreds more. Some 2,500 most-used integral solutions are given in *Table of Indefinite Integrals* by G. Petit Bois, Dover, 1961.

A list of over 10,000 closed-form solutions to integrals are available in *Table of Integrals, Series, and Products* by Gradshteyn and Ryzhik, Academic Press, 1965. While attempts are being made to teach the computer to find new "closed form" solutions ("symbolic" integration), programs are now available that find known closed form solutions. The knowledge does not yet exist which tells if a closed form solution is impossible. At least then we would know if searching would be in vain.

Incidentally, when the computer came into being, on checking the closed-form solutions given in math books, it was found that 10% of all the solutions were wrong. For a hundred years engineers were unknowingly using incorrect equations!

**Table 2.1** Sample "Table of Integrals"

| $y$ form | $\int y dx$ |
|---|---|
| 1. $ax^n$ | $ax^{n+1}/n+1$ where $n \neq -1$ |
| 2. $a/x$ | $a \ln x$; ($\ln = \log_e$) |
| 3. $(ax+b)^{-1}$ | $(1/a)\ln(ax+b)$ |
| 4. $1/(x^2+a^2)$ | $(1/a)\tan^{-1}(x/a)$ |
| 5. $1/(x^2-a^2)$ | $(1/2a)\ln(x-a)/(x+a)$ |
| 6. $e^x$ | $e^x$ (Notice this peculiar function. It can represent |
| etc. | occurrences in nature more than any other function.) |

## Practice problem

For the parabola $y = 4x - x^2$ (Figure 30), find the area between the limits $x = 0$ and $x = 4$.

**Step 1** We sketch the curve by finding several points between $x = 0$ and $x = 4$. We are verifying that the entire curve lies above the $x$-axis.

**Step 2** We notice that the area we are interested in is all above the $x$-axis, meaning the area will all be positive. If part of the curve had been underneath the $x$-axis, then we would evaluate each area separately, since the area below the axis will come out negative. This is why the area should be sketched.

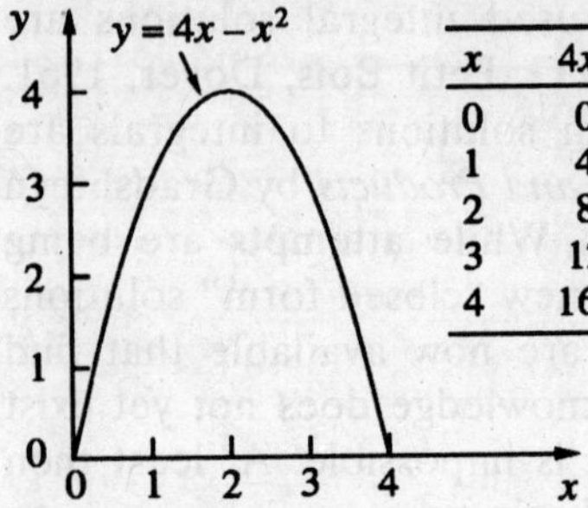


| $x$ | $4x - x^2$ | $y$ |
|---|---|---|
| 0 | 0 − 0 | 0 |
| 1 | 4 − 1 | 3 |
| 2 | 8 − 4 | 4 |
| 3 | 12 − 9 | 3 |
| 4 | 16 − 16 | 0 |

**Figure 30** *Sketch (partial) of a parabola.*

**Step 3** Sketch an element of area that shows the principle of integration—whose height is $y$ and width is $dx$ (Figure 31). The element can be anywhere within the curve, it is only for the convenience in visualizing the addition of them. Each element of area ($dA$) is the product of length $y$ and width $dx$. That is, $dA = ydx$. Of course, the width $dx$ of the element is very greatly widened for illustration purposes.

**Step 4** Now, before you integrate the expression $dA = ydx$, substitute for $y$ from the original expression $y = 4x - x^2$; we then get $dA = (4x - x^2)dx = 4xdx - x^2dx$.

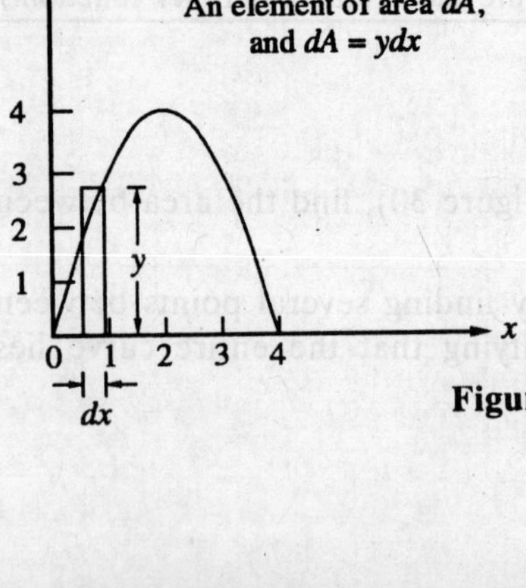


**Figure 31** *The parabola* $y = 4x - x^2$.

We could also separate each term if we want to.

**STEP 5** Now place the integral signs in their proper places; also, the limits of integration.

$$\int_0^4 dA = \int_0^4 4x\,dx - \int_0^4 x^2\,dx.$$

This indicates that we will sum up all the elements with an unlimited number of very, very thin rectangles of height $y$. (The same process, just as previously shown.) The integral of $dA$ (i.e., $\int dA$), is $A$.

**STEP 6** Now integrate each term:

$$\text{Area} \quad A = \left[\frac{4x^2}{2} - \frac{x^3}{3}\right]_0^4. \quad \text{(between } x = 0 \text{ and } x = 4\text{)}$$

**STEP 7** Now substitute the upper limit 4 for $x$.

$$A = 4(4)^2/2 - (4)^3/3.$$

Do the same thing separately for the lower limit, which is 0.

**STEP 8** After substituting the lower limit 0 for $x$, subtract those terms from the previous terms. Always use brackets on both sides of the minus sign.

$$A = \left[\frac{4(4)^2}{2} - \frac{(4)^3}{3}\right] - \left[\frac{4(0)^2}{2} - \frac{(0)^3}{3}\right]$$

$$= \left[\frac{64}{2} - \frac{64}{3}\right] - 0$$

$$= 32 - 21\tfrac{1}{3}$$

$$= 10\tfrac{2}{3}. \quad \text{Answer}$$

Notice, also, that we can integrate without knowing the limits.

This is called an "indefinite integral". For example, the function

$$y = 4x - x^2$$

integrates into

$$\text{Area} = 2x^2 - x^3/3 + C.$$

The $C$ stands for a constant whose numerical value we do not know yet, but which indicates all possibilities of the limits. When we reverse the process (i.e., differentiating or slope-finding), the $C$ drops out and we end up with the original function $4x - x^2$, this is the only true test that we have integrated correctly. (Appendix G shows why the $C$ drops out when the limits are known.)

EXAMPLE

Evaluate the following integral:

$$\int_0^1 \underbrace{(x^2 - 2x + 3)\,dx}.$$

This entire term to be integrated is called the "integrand"

SOLUTION

First integrating term by term,

$$\left[\frac{x^3}{3} - \frac{2x^2}{2} + 3x\right]_0^1.$$

Here we do not end up with $C$ since this is a definite integral (i.e., the limits are known).

## A check of integration

We can check if we integrated correctly by reversing, or differentiating term by term:

$$\frac{d}{dx}\left[\frac{x^3}{3} - \frac{2x^2}{2} + 3x\right];$$

the $d/dx$ means we want to differentiate the terms inside the parenthesis.

We now get this; $(3x^2/3 - 4x/2 + 3)$, which is the same as $x^2 - 2x + 3$, which is identical to the integrand that we started with. (When we integrate $x$, which is $x^1$, we use this exponent to multiply the coefficient of $x$, but then we subtract 1 from the exponent ($x^{1-1}$). We recall that $x^0 = 1$. Thus only the coefficient of $3x$, i.e., 3, remains.)

But continuing the problem, we first substitute 1 (the upper limit) for $x$, then 0 (the lower limit) for $x$, and subtracting the latter from the former;

$$\left[\frac{(1)^3}{3} - \frac{2(1)^2}{2} + 3(1)\right] - \left[\frac{(0)^3}{3} - \frac{2(0)^2}{2} + 3(0)\right].$$

Be careful not to change any of the signs within the second bracket, this is the most frequent careless error. In this case, the second bracket drops out anyway due to the zeroes in the numerators, giving us

$$\left[\frac{1}{3} - 1 + 3\right] = \left[2\tfrac{1}{3}\right].$$

## Some practice problems of integration (Area-finding)

A certain amount of practice in manipulation is helpful. A few problems and their answers are given on the next page. (See Table 2.1 for integrating different forms of equations.)

There are now in the market low-priced hand calculators and computers that can automatically calculate a definite integral to any desired accuracy, and can also do other things like searching and finding roots of equations.

Appendix E contains an integration program in BASIC, which will find the area under a curve to any desired degree of accuracy.

| Problem | Answer |
|---|---|
| 1. $\int_0^1 x^2 dx$ | $\frac{1}{3}$ |
| 2. $\int_2^3 x^3 dx$ | $\frac{65}{4}$ |
| 3. $\int_2^5 (2x-3)dx$ | 12 |
| 4. $\int_1^4 \sqrt{x}dx$ (note that $\sqrt{x} = x^{1/2}$) | $\frac{14}{3}$ |
| 5. $\int_{-1}^2 (t^2+t+1)dt$ | $7\frac{1}{2}$ |
| 6. $\int_2^5 \frac{2}{7x^2}dx$ $\left(\text{note that } \frac{1}{x^2} = \frac{x^{-2}}{1}\right)$ | $\frac{3}{35}$ |
| 7. $\int_2^3 \frac{5}{x}dx$ $\left(\text{note that } \int \frac{dx}{x} = \ln x + C \text{ and ln is the abbreviated form of } \log_e\right)$ | 2.027 |
| 8. $\int \frac{3dx}{(x^2-a^2)}$ | $\frac{3}{2a}\ln\left[\frac{x-a}{x+a}\right] + C$ |
| (Since this is an indefinite integral (no limits), there is no definite numerical answer.) See formula 5 in Table 2.1. | |
| 9. $\int_1^2 e^x dx$ (note $\int e^x dx = e^x + C$) | $e^2 - e^1 = 4.67$ |
| 10. $\int (x^3+5x^2-4)/x^2 dx$ | $\frac{x^2}{2} + 5x + \frac{4}{x} + C$ |

## Applications of area-finding

Applications of area-finding (integration) or (anti-differentiation), are made possible because a very large number of problems can be transformed into mathematical language and symbols, and can be exactly interpreted as *areas*. (Just as numbers count things, lines represent or measure things.) The following is only a very small, but representative list of uses. More and more are found all the time.

### FOR ENGINEERING

1. The area under a curve, to find the weight of curved objects before manufacturing them, or just the size.
2. The exact length of a curve.
3. The center of gravity of an area as well as the centroid of a solid. (The same principle for two dimensions applies to three dimensions.)
4. The center of percussion, the center of pressure, torque, the force exerted on a submerged body.
5. The volume of an area rotated about any axis.
6. The surface area of a hollow solid of revolution.
7. Work done, energy, moments of inertia.
8. Many many other quantities as well.

### FOR BUSINESS AND ECONOMICS

1. Total cost of producing and marketing.
2. Revenue and demand functions.
3. Marginal revenue to consume, to save.
4. Consumer's surplus, producer's surplus.
5. Total net earnings.
6. Maximising profit, minimising cost, minimising taxes, etc.
7. Capital value.

### FOR PROBABILITY AND STATISTICS

1. Probability of certain outcomes.
2. Probability density functions.

FOR BIOLOGY, SOCIAL SCIENCES, ETC.

1. Natural growth and decay (for example: population growths, bacteria growths).
2. Carbon dating, etc.

The number of uses in the sciences, physical and non-physical, is virtually without limit.

## More than just area-finding

Integration, i.e. area-finding's main and most valuable contribution is in *function* (formula) *finding.* Apart from guessing or from the insight of a genius, there is no other way of finding the formulas that describe complicated problems other than by the rate of change. This is because it is far easier to describe physical situations in terms of rates of change (the derivative) than by any other means. Then the rate of change within the formulas must be changed into a simpler and useful form by integration. A computer can find the numerical answer to any area, it can also be made to find the slopes at particular points on a line, but it cannot be made to find a mathematical formulation of a physical law. But the use of calculus can do it by the use of the *Differential Equation.* It is conceivable that one day the computer will be able to formulate physical laws in mathematical form. When that happens, mankind will rocket ahead in physical progress beyond anything dreamt till now.

# 3

# The Wedding of Slope-Finding and Area-Finding

## Differentiation, Integration and its Resulting Brainchild—The Differential Equation

This is the place where all the action is! The application resulting from the use of the Differential Equation is absolutely staggering. There is virtually no field of human endeavour that has not been and will not still be further transformed forever by the application of this brainchild. And one day, not far from now, we will travel to other stars, reach the extremities of the universe, mix with people from other worlds, and find the secret of eternal life. (This is not farfetched at all.) I believe aging is merely an automatic, or accidental "destruct" mechanism built into us. We need only discover the mechanism and dismantle it. The solution could be under our very noses. All of this and more, no small thanks to these very methods of analysis.

Here is an example that shows how by using differentiation (slope-finding) and integration (area-finding), we can find the mathematical formula that describes to perfection a certain physical event. This particular example is not earth-shaking in itself, but we are now looking for the principles involved which will lead us to be able to predict and thereby control events. This example, however, can be checked easily by yourself without the use of expensive equipment. Essentially, all you need is a thermometer and a wristwatch.

Do not be concerned if you cannot now follow every single piece of mathematics that follows. We are demonstrating the truth of our previous claims.

## Example for formulating a differential equation

I take a thermometer in from the cold outdoors (it is now winter outside, 37°F) and bring it indoors, which is 76°F. Can we express with a *formula* the temperature of the thermometer with respect to time as it warms up? For example, what will the thermometer temperature be after 1 minute? After 95 seconds? (of course the same principle applies if we bring in an object from any medium where it is *warmer* into any cold medium.) Maybe there is no such formula? Maybe each time you do exactly the same thing there occurs a different result like the throw of a dice? How can we tell? You should verify this example yourself and you only need a thermometer that shows individual degrees (if possible) or that has a large number of "markings" per unit length. You do your own experiment, and while the thermometer is reading the indoor temperature ($F$), write it down first, only then put the thermometer outdoors and let it settle to the outdoor temperature ($T$). Write that down. When you've got a watch with a second hand and a pencil and paper ready and the temperature *is settled on the thermometer, bring in the thermometer* from the outdoors (it is best to do it from your window, but don't allow the sunlight on the thermometer because that introduces a complication which we don't need now). Exactly at that instant when you bring in the thermometer from the outside, the time countdown starts at $t = 0$ minutes. Now lay down the thermometer and record the temperature at half minute intervals for 5 minutes or longer. (Do not breathe on or handle the thermometer or introduce any other temperature influences on it except for the room temperature which you have already measured.) By "room temperature" we are, of course, referring to the temperature of the air in the part of the room you are doing this.

After marking down the temperature as accurately as possible, in 30 second intervals, graph it with time as the horizontal axis (time is conventionally graphed as the horizontal axis). Here is the data I collected:

| $t$, Time (min.) | 0 | .5 | 1 | 1.5 | 2 | 2.5 | 3 | 3.5 | 4 | 4.5 | 5 |
|---|---|---|---|---|---|---|---|---|---|---|---|
| $T$, Temp. (°F) | 37 | 44 | 47 | 51 | 54.5 | 58 | 60 | 62.5 | 65 | 67 | 68 |

This is how it will look (Figure 32) on a graph. I myself am using a Woolworth's thermometer, which is about the only one I could easily find. It is graduated in 5 degree intervals of Fahrenheit, and one inch covers 60°, which is certainly not the most ideal one. If possible try to find one with fewer degrees per inch for reading ease and greater accuracy. If it is in centigrade and centimeters, the same procedure follows. Take care to read the temperature every 30 seconds with as much accuracy as you can, even to half a degree if possible.

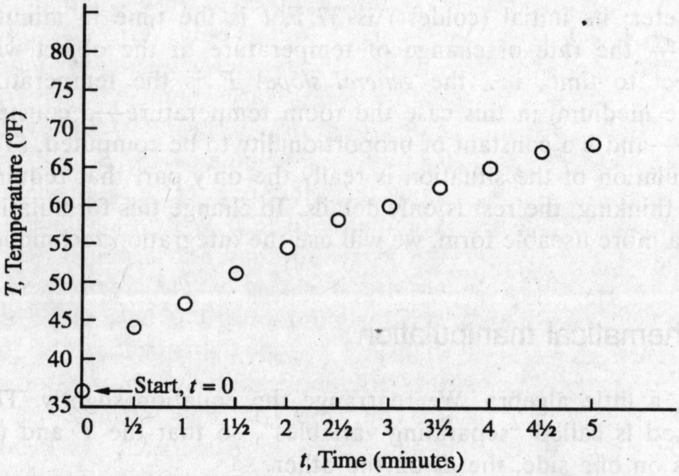


**Figure 32** *Experimental data for the changing temperature.*

We will call this the "Experimental Data" because it is not theory, but reality. In the mathematical way we will call it "Theoretical" for now.

Now let us do it the mathematical way, using our new found ways: we start with what is called "Newton's Law of Cooling" (it applies to heating as well). This says that if an object (the thermometer from the outdoor temperature) is immersed in a medium (indoor room air temperature) and if the medium is kept at a constant temperature, then the temperature of the object (the thermometer) will change at a rate proportional to the difference in temperature between the object (the thermometer) and the medium (indoor temperature). It is the same thing as saying for example, that if you have a fire, the hotter it is, the more quickly will it heat up something. Incidentally, this very same formulation is responsible for a whole slew of scientific laws. Mathematically, this problem is expressed this way (do not confuse the letters $t$ (time) with $T$ (temperature)):

$$\frac{dT}{dt} = \overbrace{k(T - F)}^{\text{Difference in temp.}}.$$

Mathematical formulation of ''Newton's law of cooling''

Here, $T$ is the temperature of the object, in this case the thermometer; its initial (coldest) is 37°F, $t$ is the time in minutes and $\frac{dT}{dt}$ the rate of change of temperature of the object with respect to time, i.e., the *general slope*! $F$ is the temperature of the medium, in this case the room temperature—a constant 76°F—and $k$ a constant of proportionality to be computed. (The formulation of the situation is really the only part that requires hard thinking, the rest is only details. To change this formulation into a more useable form, we will use the integration technique.)

## Mathematical manipulation

First, a little algebra. We rearrange the equation slightly. This method is called "separating variables", so that the $T$ and the $dT$ is on one side, the $dt$ on the other.

$$\frac{dT}{(T - F)} = kdt.$$

Now we anti-differentiate (integrate) both sides (we can do anything we want to, to one side of an equation if we do it to the other side as well). The object is to transform the $dT$ and $dt$. In this case area-finding is not the prime object of this exercise, it is the anti-differentiation quality that we want, so that we can have a *useful equation.*

*This is because it seems to be very much easier to describe physical events in terms of the slope (rate of change) as in this case, than by any other way. This is also why we needed more than just the actual numerical area under a curve—a general integration method for any curve.*

We perform an integration on both sides of the equation

$$\int \frac{dT}{(T-F)} = \int kdt,$$

which then becomes,

$$\ln(T-F) = kt + C.$$

← The $C$ is used when limits are not specified and applies to both sides

↑ See formula 2, Table 2.1 (under $\ln(T-F)$)

↑ The result of the integration of the constant $k$ (under $kt$)

If we now bring each side to the power of $e$, then ln disappears, since $e^{(\ln x)} = x$. This is a very useful identity to be memorized. We now have

$$e^{\ln(T-F)} = e^{kt+C}.$$ We are treating each side equally

Using $e^{\ln(T-F)} = (T-F)$,

$$T - F = e^{kt+C} = e^{kt}e^{C}.$$

Since $e$ and $C$ are each constants, we might just as well put the constants together, and the symbol $C$ need not be changed.

$$T - F = Ce^{kt},$$

or $\quad T = F + Ce^{kt}.$ ← This is the general formula

Now for the specific formula we have to evaluate $k$ and $C$ which we can do firstly by taking into consideration the "initial condition" that when $t = 0$, that is at the instance that the thermometer was brought into the house, the temperature ($T$) was equal to 37°F. Also, when we let $t = 0$, it is very convenient because

$$e^{k(0)} = e^0 = 1.$$ Anything (except zero) to the zero power = 1

This enormously simplifies the equation. Inserting also $T = 37$, and $F = 76$ so that, $37 = 76 + C$ or $C = -39$, gives us

$$T = 76 - 39e^{kt}.$$

In order for this equation to be useful to us we must still evaluate the other unknown $k$. For example, looking at the data at $t = 2$ minutes, we have $T = 54.5°$F. (The same outcome would occur with any other data point.) Now we substitute for them both at the same time.

$$54.5 = 76 - 39e^{k(2)} \quad \text{or} \quad -21.5 = -39e^{2k}$$

$$0.551 = e^{2k}$$

$$2k = \ln 0.551.$$ We again take the ln of both sides (identity: $\ln e^x = x$), reversing too

Either use a pocket calculator to find $\ln 0.551$ or a table of natural logarithms, so

$$2k = -0.596$$

$$k = -0.298.$$

Now we have the full equation with all the pieces described. Putting them all together, including the $k$ value,

$$T = 76 - 39e^{-.298t}.$$

*This is the completed specific formula!*

This equation gives the temperature of the thermometer at any time after it has been brought in from the outside 37°F into the room which is 76°F. A few checks are now in order (Figure 33). Let's try $t = 3$ minutes in the formula to see how it compares with the experimental data. Then,

$$T = 76 - 39e^{-0.298(3)} = 76 - 39(0.409)$$

$$= 60° \text{ F.}$$ This hits the mark on the head!

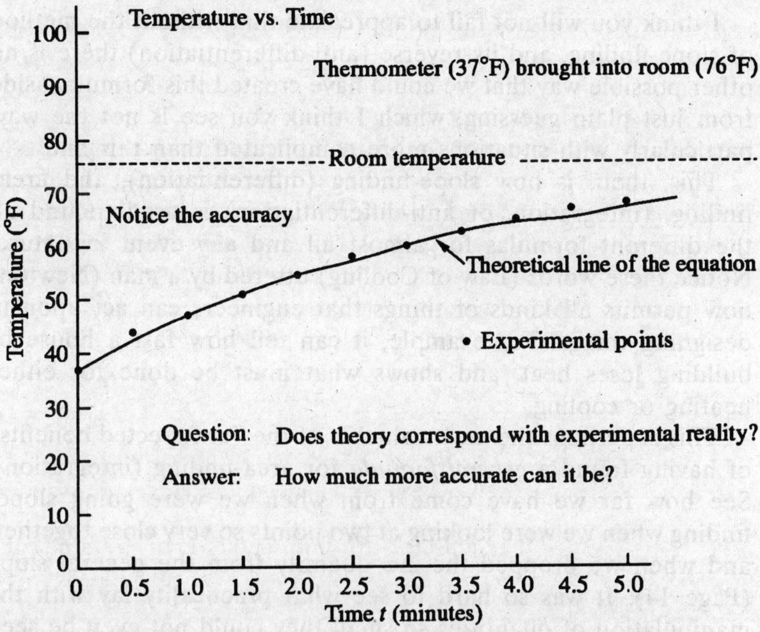


| Time $t$ (minutes) | 0 | 0.5 | 1 | 1.5 | 2 | 2.5 | 3 | 3.5 | 4 | 4.5 | 5 |
|---|---|---|---|---|---|---|---|---|---|---|---|
| Experimental(°F) | 37 | 44 | 47 | 51 | 54.5 | 58 | 60 | 62.5 | 65 | 67 | 68 |
| Theoretical(°F) | 37 | 42.4 | 47 | 51 | 54.5 | 57.4 | 60 | 62.5 | 64.1 | 65.8 | 67.2 |

**Figure 33** *Comparison of theory to experiment.*

We have now created a formula which can be graphed as a curve! To check with another value, try $t = 5$ minutes, then

$$T = 76 - 39e^{0.298(5)} = 76 - 39(0.2254)$$

$$= 67.2° \text{ F.}$$ This one is almost exactly on the head. I'm sure that you get the idea!

Look at the graph in Figure 33. See how close the experiment agrees with theory! I had to repeat the experiment several times to get the timing right and the writing of the temperature at the same time. You can see that the simple act of bringing something in from the cold follows a precise law which will not fail no matter how many times you try it. The temperature change is neither accidental nor random. Not at all!

I think you will not fail to appreciate that without the method of slope-finding, and its reverse (anti-differentiation) there is no other possible way that we could have created this formula, aside from just plain guessing, which I think you see is not the way, particularly with situations more complicated than this one!

This, then, is how slope-finding (differentiation), and area-finding, (integration, or anti-differentiation) is used to find all the different formulas for almost all and any event in nature. Notice these words (Law of Cooling) uttered by a man (Newton) now permits all kinds of things that engineers can act upon in designing things. For example, it can tell how fast a house or building loses heat, and shows what must be done for either heating or cooling.

This is perhaps the most valuable of the "unexpected benefits" of having found a *general formula* for area-finding (integration). See how far we have come from when we were going slope-finding when we were looking at two points so very close together, and when we dropped the $\Delta x$ quantity from the general slope (Page 14). It was so hard to see what practicality lay with the manipulation of quantities so small they could not even be seen with the eye! It looked as if we were splitting hairs for no useful purpose! Remember also when we were listing the advantages of being able to automatically find the slope from a given function (Page 18). At the time it hardly seemed worth the effort, and we were hard put to list the advantages that would come with it!

Here is one more example of how an equation is derived by using slope-finding, and its reverse (area-finding, or anti-differentiation, or integration) technique. We will derive the equation governing another natural law. To the inventors of natural laws, this would be an excellent way of testing the laws out.

## The absorption law of Lambert

*Statement*: The rate of change of the amount of light ($L$) in a thin transparent layer with respect to the thickness ($h$) is proportional to the amount of light on the layer.

When light passes through a transparent medium (like a pane of glass, or a layer of water), the amount absorbed by any thin layer of the material perpendicular to the direction of the light is proportional to the amount incident on that layer and the thickness of the layer. (The thicker the layer the more is absorbed, i.e., less light goes through. Also the stronger the light, the more light is absorbed and also passes through.) Find the formula of this law. (Notice that you don't have to know anything about Physics or Light to be able to formulate this law with confidence. Many other physical laws have reasoning that differs little from the above. Some of these fields are gravity, electricity, magnetism, decay, growth, etc.)

SOLUTION

$\frac{dL}{dh} = kL.$ Mathematical formulation of this law, with the derivative

Here $\frac{dL}{dh}$ is the rate of change of $L$ with respect to $h$.

We divide both sides of the equation by $L$, then multiply both sides by $dh$, just for separating the variables.

$\frac{dL}{L} = kdh$ Now we are ready to integrate. We do so to both sides

$$\int \frac{dL}{L} = \int kdh.$$

Then $\ln L = kh + C.$ Bring each side to the power of $e$

$e^{\ln L} = e^{kh+C},$ From the identity $e^{\ln x} = x$

$$L = e^{kh+C},$$

or, $L = e^{kh}e^{C}.$ Note this algebraic rule: $a^{x+y} = a^x \cdot a^y$

As before, let $e^C = C$ because both $e$ and $C$ are constants anyway, then,

$L = Ce^{kh}.$ This is the general formula

We could add another small refinement without going into any detail, because all we want to do now is prove the point!

If we wanted to evaluate $C$ and $k$, we know that when $h = 0$, then $L = C$. A way that is used to show that $C$ is the *initial condition* of $L$ is by designating it $L_0$. Replacing,

$L = L_0 e^{kh}$. `The slight refinement has been added!`

$L_0$ means that amount of light on a layer of zero thickness (which is the technical way of saying no thickness, or nothing at all). Thus the quantity of light is measured before a layer of transparency is put down. To actually evaluate $L_0$ and $k$, we only need the actual experimental data. But we already know the general shape of the curve and we already have a "feel" for what will happen as we thicken the layer.

*This is how Lambert may have discovered his law of absorption of light by experiment, not by mathematical formulation, and how you can perhaps make a discovery of your own law.*

1. Supposing he had a light intensity measuring device, a light meter. He could have, say, a half dozen panes of translucent material, all of the same composition and the same thickness.

2. Taking, the light meter, he would place it on the floor, under a source of light (such as the ceiling light). He records the light meter reading with no translucent panes on it. Then he places one pane over the light-sensitive part (the photo cells) of the meter and jots down the new reading. Of course, the intensity of light will read lower because the translucent material is absorbing some of the light. Then he places a second pane on top of it and takes the new reading. The light intensity felt by the photo cells will be even less because more light will be absorbed by the two translucent panes. Then a third and fourth and fifth, etc., pane is piled on and a reading taken for each. It is clear that the light intensity measurement is a measure of the absorption of the panes; i.e., the thicker the pile of panes, the lesser the light that goes through because more light is being absorbed. It goes without saying that the more the light that is shining

(move the light meter closer to the light source), the more the light that will be absorbed, and also the more the light that will manage to exit through the translucent panes.

3. Now Mr. Lambert makes a picture of the situation (Figure 34).

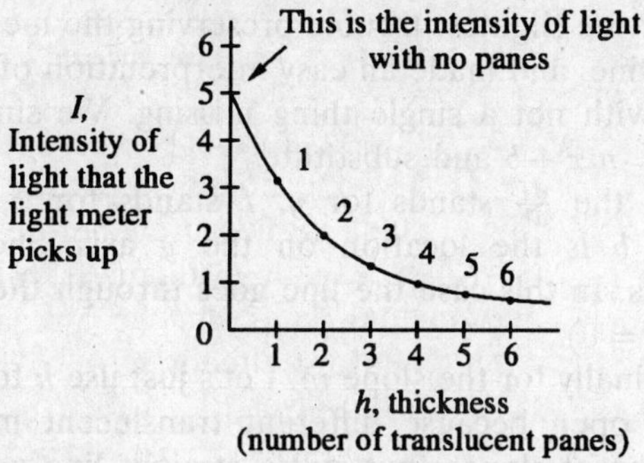


**Figure 34** The absorption of light vs. translucent layers.

4. He does not know what to do with the data curve in the above figure. He can try to fit an artificial formula through the points (i.e., make up a formula), but would rather not, since this artificial formula, even though it might fit very well, will cast no real meaning on the situation. So he tries every kind of mathematical contortion he can think of, and finally stumbles on the following:

5. Since, from Figure 34, it is possible to measure the slope (which is $dI/dh$) with a ruler, at each point along the curve he records the slope vs. intensity ($I$) from Figure 34. He makes a new graph, like the one in Figure 35.

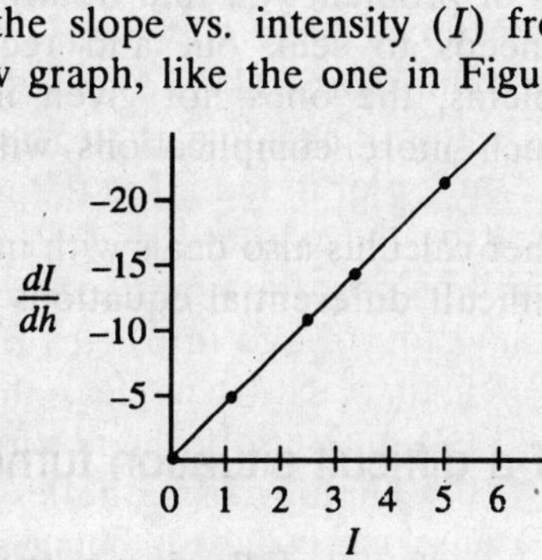


**Figure 35** *Slope vs. intensity.*

6. He has struck it exceedingly lucky! (Remember $y = mx + b$ is the form of equation for every straight line.) This means that from the data of the first curved line ($I$ vs. $h$) he has obtained a straight line ($dI/dh$ vs. $I$). There is no problem in finding the exact equation for any straight line. Also, we have not the slightest trouble preserving the meaning of this straight line, and made an easy interpretation of every single symbol with not a single thing missing. We simply use the form $y = mx + b$ and substitute.

   Here, the $\frac{dI}{dh}$ stands for $y$, $I$ stands for $x$, and $b = 0$, because $b$ is the location on the $y$ axis where the line intersects. In this case the line goes through the origin (i.e., where $y = 0$).

   And finally for the slope $m$. Let's just use $k$ for it for now. Keep it open because different translucent materials will give different slopes (but still a straight line going through the origin). So putting it all together, and substituting into the straight-line form of $y = mx + b$, we get

$$\frac{dI}{dh} = kI. \qquad \texttt{Lambert's law of the absorption of light}$$

7. Now of course this *is* Lambert's Law! He found it!

This is what creative mathematics and science is about: Keep translating things that are difficult (the curved-line data) into things that are easy (the straight line). This is almost like problem-solving, of the kind of problem you find in any textbook, except that at first one needs to seek out and recognise what are the unknown problems, the ones not given in any textbooks. Situations with much more complications will also work out similarly.

Much of the higher calculus also deals with ingenious methods of reducing very difficult differential equations into simple ones like the following:

## One example of a difficult situation turned simple

Do not concern yourself with following every single step that follows. Would you believe (for example) that this really nasty looking equation called the Cauchy or Euler equation

$$x^2\frac{d^2y}{dx^2} + ax\frac{dy}{dx} + by = 0$$

(where $\frac{d^2y}{dx^2}$ is the symbol for the second derivative) can be easily solved (the object is to find $y$) when treated as though it were an *algebraic equation*? (Can you believe that?)

SOLUTION

If we let $x = y^m$, take its derivatives, etc., and substitute, we get

$$x^2m(m-1)x^{m-2} + axmx^{m-1} + bx^m = 0.$$

The above looks formidable, but it is an *algebraic equation* (with no derivatives). This is solved to be $y = C_1x^{m_1} + C_2x^{m_2}$, where $m_1$ and $m_2$ depend on $a$ and $b$. (For more details see *Advanced Engineering Mathematics,* 3rd Edition, E. Kreyszig, 1969, Wiley & Sons.)

Other types of equations have equally simplifying techniques. Notice that since algebraic equations can be easily solved using calculators and computers, it follows that so can differential equations (but only for numerical answers, not for formulas).

There always seems to be an easy way to do a hard problem. But you have to know it. There are books on differential equations that will give various methods and procedures for many physical situations.

Just to show you that nature has been waiting a long time for us mortals to catch up with her, look at Figure 36. The question is, is this a picture of

- a knot of wood within a tree?
- flow of water around an obstacle?
- a mathematical diagram made from the very methods we have used?

I am sure we know it is all the three, identically!

Again, it is not only the knot of the tree that we are talking about, nor the flow of water around a circular obstacle, it is the

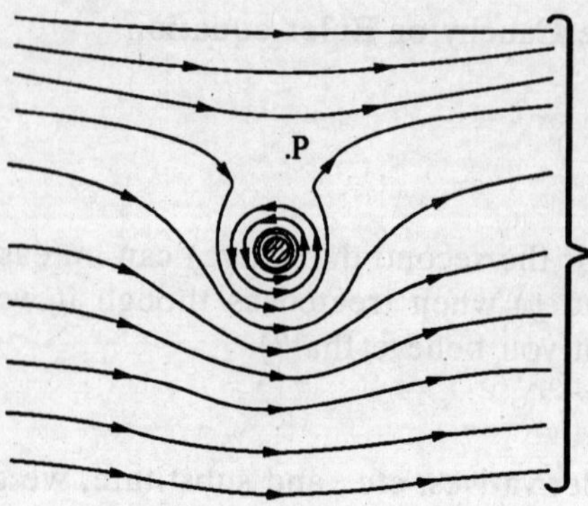


**Figure 36** *The Laplace equation.*

principle of the situation! You can bet that not only all physical phenomena are perfectly describable with differential equations, but all living things as well. But most of the formulas have yet to be discovered. Perhaps you can be of help?

THE END – Or is it just the beginning for you??

---

[1] The signs $\Delta$ and $\partial$ are just other ways of expressing $dy/dx$, the slope, when there is more than one variable on the right side of the equation.

# Appendix A

PROBLEM 1

The government has to fence in 60,000 square feet of land in a rectangular plot along a straight highway. Fence costs Rs 1 per running foot along the highway, and the other three sides cost Rs 0.75 per foot. How much of each fence is necessary for minimum cost? What is this minimum cost? What are the dimensions?

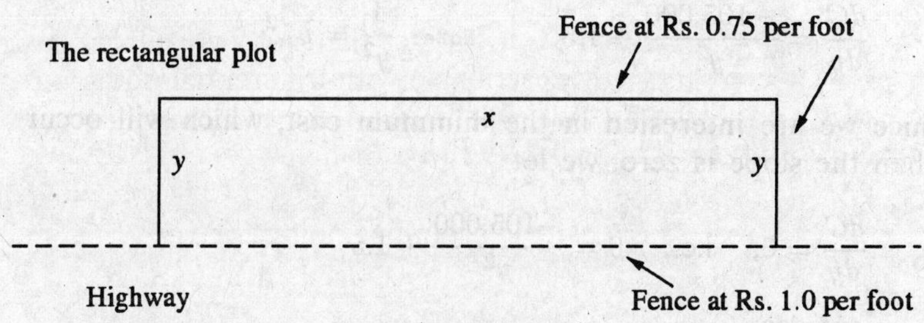


Area $A = xy$ (for all rectangles).

Cost of $x$ length near highway = (cost/length) (length) = $1x$.

Cost of the other $x$ length = $0.75x$.

Cost of one $y$ length = $0.75y$.

$$\text{Cost } C \text{ of the entire fence} = 0.75y + 0.75y + 1 \cdot x + 0.75x$$

$$= 1.75x + 1.5y.$$

Now we have got the thing which we want to minimize (cost $C$, in this instance) alone on the left hand side. Before we differentiate (that is, find the slope), we must have only one unknown (the independent variable) on the right-hand side. We can either replace $x$, or we can replace $y$, by using the equation $A = xy$; since it is given that $A$ = 60,000, then 60,000 = $xy$.

Let us solve for $y$ this time. We divide both sides by $y$, getting $x = 60000/y$, which we substitute into the cost equation

$$C = 1.75\left(\frac{60{,}000}{y}\right) + 1.5y.$$

This is the same as

$$C = 105{,}000y^{-1} + 1.5y.$$

Since the cost $C$ is alone on the left-hand side of the equation, and only one unknown is on the right-hand side, we will do our slope-finding procedure, which is also called differentiating:

$$\frac{dC}{dy} = \frac{-105{,}000}{y^2} + 1.5. \qquad \texttt{Note:}\ \frac{1}{y^2} = y^{-2}$$

Since we are interested in the minimum cost, which will occur when the slope is zero, we let

$$\frac{dC}{dy} = 0; \quad \text{i.e.,} \quad 0 = \frac{-105{,}000}{y^2} + 1.5$$

$$\frac{105{,}000}{y^2} = 1.5 \quad \text{or} \quad 1.5y^2 = 105{,}000 \quad \text{or} \quad y^2 = 70{,}000.$$

Ignore the negative part of the answer, and so, $y = +264.58$ feet.

To obtain $x$ we merely substitute into the nearest equation that has $x$ and $y$ in it, i.e., $60{,}000 = xy$ the value $y = 264.58$ ft. to obtain $x = 60{,}000/264.58 = 226.78$ ft. The actual minimum cost is easily obtained by substituting both $x$ and $y$ into any of the cost equations, say, $C = 1.75x + 1.5y$.

$$C = 1.75(226.78) + 1.5(264.58)$$

$$= \text{Rs. } 793.74. \qquad \texttt{Answer}$$

You may check, but you will not be able to find other dimensions for the rectangle which will cost less. Note too, that this problem could have been solved by trial and error as well. Slope-finding here is actually not absolutely essential, but it is definitely a labour-saving device!

PROBLEM 2

A farmer wants to fence in 200,000 square feet of land in a rectangular plot, then divide and fence that plot into three equal plots. Find the least amount of fence necessary to do this.

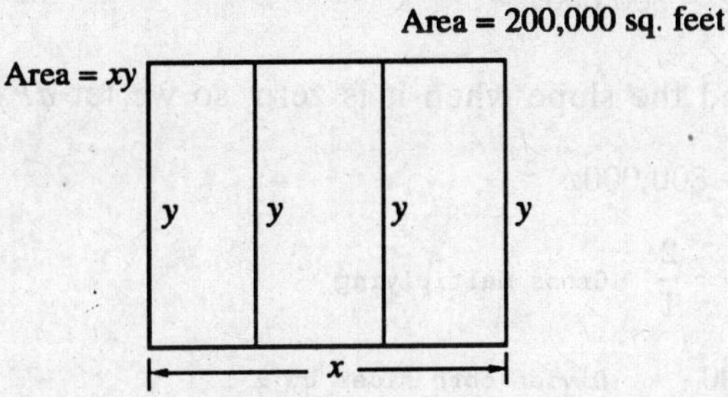


Since Area $A$ = 200,000 and also $A = xy$, then 200,000 = $xy$. Now since, in this problem we wish to minimize the amount of fence, we must first describe by mathematical symbols the amount of fence on the left-hand side of the equation equal to one unknown, either $x$ or $y$. Always, the thing you wish to minimize (or maximize) must be alone on the left-hand side of the equation, and it should be pictured in your mind as representing the vertical axis of a graph, while the one unknown you are going to test out (either $x$ or $y$ in this case), must be on the right-hand side of the equation. That would be represented by the horizontal axis. Let us then symbolize the entire length of fence as $F$. Now adding up all the lengths of fence we have,

$$F = 2x + 4y.$$

Again, solving for $x$ or $y$ (this time for $y$, for instance),

$$y = \frac{200{,}000}{x},$$

and substituting it into the length of fence equation,

$$F = 2x + \frac{4(200{,}000)}{x} = 2x + 800{,}000x^{-1}.$$

Now you see $F$, the thing you wish to find, alone on the left hand side of the equation, while only one unknown, in this case

$x$, is on the right side of the equation. Now we are ready to find the slope! Again, we wish to find the location of $x$ where the slope is minimum, so we find the slope (differentiate).

$$\frac{dF}{dx} = 2 - 800{,}000x^{-2}.$$

We only need the slope when it is zero, so we let $dF/dx = 0$.

$$0 = 2 - 800{,}000x^{-2},$$

or $\dfrac{800{,}000}{x^2} = \dfrac{2}{1}$ `Cross multiplying`

$2x^2 = 800{,}000$ `Divide both sides by 2`

$x^2 = 400{,}000$ `Take the square root of both sides`

$x = \pm(632.46)$. `We ignore the negative answer, since we cannot have a negative length`

Substitute into the area equation to get $y$,

$$y = \frac{200{,}000}{632.46} = 316.23 \text{ feet.}$$

The actual minimum fence can easily be gotten by using any of the length-of-fence equations, one of which is $F = 2x + 4y$, where we now know both $x$ and $y$.

$$F = 2(632.46) + 4(316.23) = 2529.84 \text{ feet.}$$

This is the minimum length of fence required in this problem. Any other dimensions of $x$ and $y$ will result in more fence than this. This problem could have been done without the process of slope-finding, i.e., by trial and error, or by computer.

PROBLEM 3

Suppose there is an isosceles triangle (the two sides are 10) while the base length is not known, and you want to know what height should the triangle be such that the area of the triangle is a maximum? Also, find this area (see Figure 15).

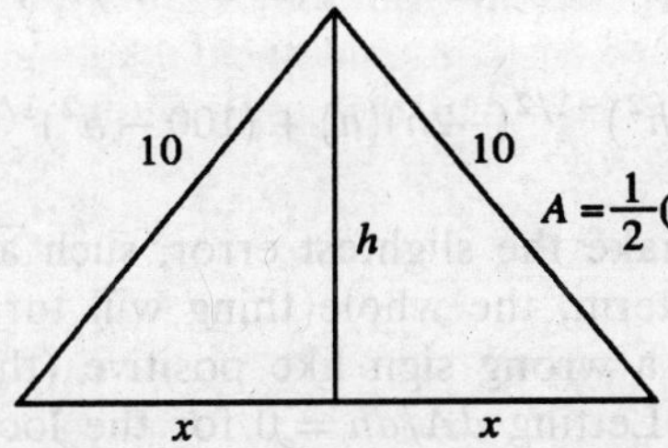


We must first find an equation of the area of this triangle. The area of every triangle is, $A = \frac{1}{2} \times$ base $\times$ altitude, or, $A = \frac{1}{2}bh$. In this problem the base $= 2x$, altitude $= h$, so

$$A = \frac{1}{2}2xh = xh.$$

In order to find the maximum area (i.e., where the slope is zero in the area vs. length curve for either $x$ or $y$), we must differentiate $A$. And to do this we must either get rid of $h$ or $x$, which means finding $h$ in terms of $x$ (or find $x$ in terms of $h$). By the theorem of Pythagoras,

$$x^2 + h^2 = 100, \qquad x = \pm\sqrt{100 - h^2}.$$

Solving for $x$ (we could also solve for $h$, and substitute for it) — Ignore the negative part of the answer

We substitute this into the area equation, because we wish to find the location of zero slope.

$$A = (100 - h^2)^{1/2}h.$$

Now we use the Product Rule which is number 5 from the Table 1.1 of derivatives

$$\frac{d(uv)}{dx} = u\frac{dv}{dx} + v\frac{du}{dx}.$$

Using this form with $u = h$ and $v = (100 - h^2)^{1/2}$, we get

$$\frac{dA}{dh} = \frac{1}{2}(100 - h^2)^{-1/2}(-2h)(h) + (100 - h^2)^{1/2}.$$

Notice that if you make the slightest error, such as leaving out the $(h)$ of the first term, the whole thing will turn out wrong. The same thing for a wrong sign like positive (the $-2h$ term) instead of negative! Letting $dA/dh = 0$ for the location of $h$ at zero slope, we have

$$0 = \frac{1}{2}(100 - h^2)^{-1/2}(-2h)(h) + (100 - h^2)^{1/2},$$

or $\quad (2h)(h)\frac{1}{2}(100 - h^2)^{-1/2} = (100 - h^2)^{1/2}.$

Simplifying and cross multiplying,

$$\frac{(100 - h^2)^{1/2}}{1} = \frac{h^2}{(100 - h^2)^{+1/2}}.$$

(Note: $(a^{1/2})(a^{1/2}) = a^{1/2+1/2} = a^1 = a$.)

$100 - h^2 = h^2$, (Notice we add $h^2$ to both the sides)

$2h^2 = 100$ or $h^2 = 50$,

$h = \pm 7.07$. We use only the positive value of $h$

This maximum area is easily found to be, by substitution:

$A = (100 - h^2)^{1/2}h$ For $h$

$= (100 - 50)^{1/2}(7.07)$

$= 50$ square units.

With a diagram of Area vs. Altitude, $h$ can easily be drawn up (see Figure 37) using the equation:

$$A = (100 - h^2)^{1/2}h.$$

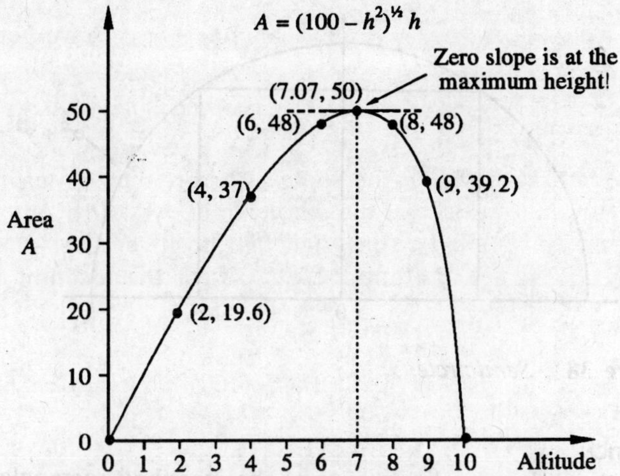


**Figure 37** *Area vs. altitude for isosceles triangle with sides 10 each.*

This problem can be done without the use of slope-finding, just by plotting the curve and picking the highest point on the curve by inspection. Of course, putting together equations like $A = (100 - h^2)^{1/2}h$, does require some practice, but it is not difficult in itself. Or one could have simply constructed various triangles and measured them using trial and error.

PROBLEM 4

Given a semicircle whose equation is $x^2 + y^2 = r^2$, find the dimensions of the rectangle with the largest possible area that can be inscribed in it, and find that area (Figure 38). After calculating the answer to this problem, graph that area of various inscribed rectangles. See if your answer checks with the result which you can get by the method of slope-finding! You can see that this problem can also be done by drawing several of these rectangles, and by trial-and-error, finding which has the maximum area. Do the problem both ways.

You will notice that it helps very much to *always sketch the problem.* Also, for all of the curves drawn, to use slope-finding you must know the equation of that curve, like that of the semicircle in Figure 38.

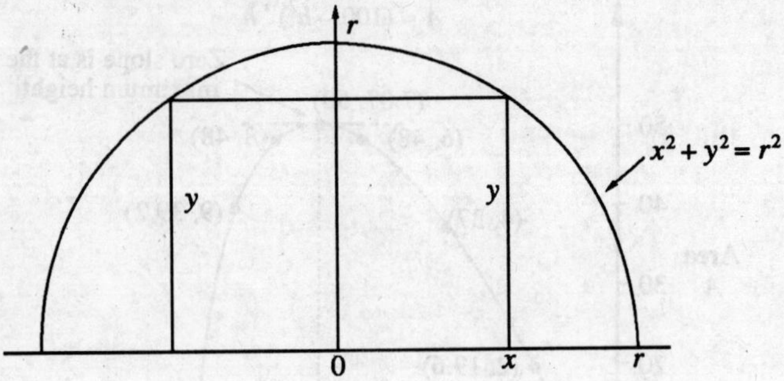


**Figure 38** *Semicircle.*

SOLUTION

Now the unknown dimensions of the inscribed rectangle we call $2x$ and $y$. The area of this rectangle must be $A = 2xy$. Since we are going to find the rectangle with the maximum area, we must use this equation because we are going to have to differentiate it. So, for $A = 2xy$, as before, we must have only one unknown on the right side of the equation. We use the equation of the circle $x^2 + y^2 = r^2$, and solve for $x$ (or for $y$). Solving for $x$ we have

$$x^2 = r^2 - y^2,$$

$$x = \pm(r^2 - y^2)^{1/2}.$$

The positive answer is sufficient, we ignore the negative answer

Now substituting in the area equation of $x$, we get

$$A = 2(r^2 - y^2)^{1/2}y.$$

(Since $r$ is a constant, it is treated as any number would be.) Now we differentiate, that is, we find the slope at all points of the curve

$$A = 2(r^2 - y^2)^{1/2}y.$$

We now have, (from the product rule, form 5 of Table 1.1, i.e.. slope-finding formulae)

$$\frac{dA}{dy} = 2\left[(r^2 - y^2)^{1/2} + \frac{y}{2}(r^2 - y^2)^{-1/2}(-2y)\right].$$

(Notice that the coefficient, 2, applies to both terms.)

Letting $\frac{dA}{dy} = 0$, and dividing both sides of the equation by 2,

$$(r^2 - y^2)^{1/2} = \frac{y^2}{(r^2 - y^2)^{1/2}}.$$

Zero divided by anything (except zero) is always zero

Cross multiplying, we have,

$$r^2 - y^2 = y^2 \quad \text{or} \quad 2y^2 = r^2$$

$$y^2 = r^2/2,$$

and finally,

$$y = \frac{r}{\sqrt{2}}.$$

The negative part of the answer is ignored

To find $x$, we substitute into

$$x^2 + \frac{r^2}{2} = r^2,$$

and we find that

$$x = \frac{r}{\sqrt{2}} \cong .707r.$$

Answer

This also means that for the area to be maximum for the inscribed rectangle in a semicircle, $x = y$, and the area $= 2xy$ is

$$A = 2\left[\frac{r}{\sqrt{2}} \cdot \frac{r}{\sqrt{2}}\right] = r^2.$$

No other dimensions of $x$ and $y$ exceed this.

PROBLEM 5

Given a sphere, find the dimensions of an inscribed cylinder which occupies as much volume as possible. Also, find this volume and compare it with that of the sphere.

SOLUTION

First we must draw the diagram, because we know we are going to have to get geometrical relationships between the unknowns. This cannot be done without the use of a sketch (Figure 39).

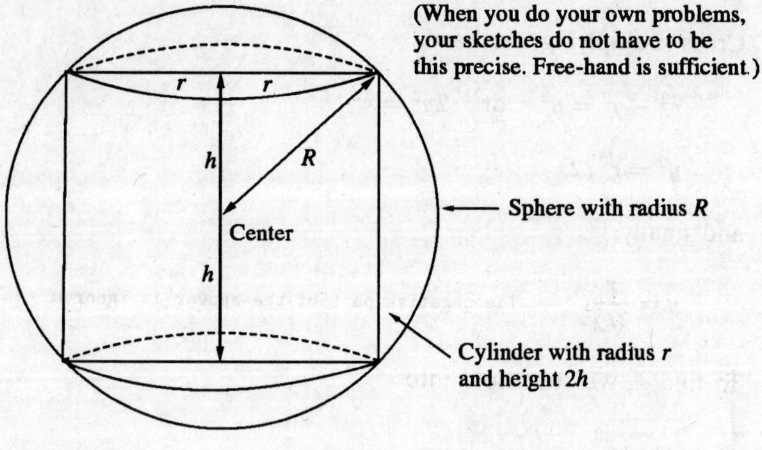


**Figure 39**

Now, since we are going to maximize the volume of the cylinder, we must have the formula for the volume of a cylinder. This is $V = \pi r^2(2h)$. The $2h$ in our problem represents the full height of the cylinder. (We could just as easily made the full height to be $h$. But using the $2h$ as the height of the cylinder helps later.)

$$V = 2\pi r^2 h.$$

We have volume alone on the left-hand side of the equation. And we have the variable $r$ and $h$ on the right side. We must get rid of one of these 2 variables by finding their relationship, and substituting for one of them. From the diagram, using the Pythagorean Theorem, we have, $h^2 + r^2 = R^2$. This is the necessary relationship, since $r$ is related to $h$, while $R$ is a constant. This will change an equation with 2 unknowns into an equation with one unknown. So we subtract $h^2$ from both sides of this equation (we know that we can do anything we want to one side of an equation, as long as we do the same to the other). So,

$$r^2 = R^2 - h^2.$$

Substituting this relationship in the volume equation,

$$V = 2\pi h R^2 - 2\pi h^3. \quad \texttt{We have got rid of the } r \texttt{ term}$$

Now, and only now, is the equation ready to be differentiated (i.e., we are now ready to find the general slope for all points of this equation). Therefore we differentiate it with respect to $h$. After all, $h$ is the only variable remaining on the right-hand side of the equation.

$$\frac{dV}{dh} = 2\pi R^2 - 6\pi h^2.$$

Since we are looking for a maximum value, we are not interested in any other point in the curve, except where the slope, $dV/dh = 0$. So we let $dV/dh = 0$.

$$0 = 2\pi R^2 - 6\pi h^2$$

or $\quad 6\pi h^2 = 2\pi R^2 \quad$ `Now divide both sides by` $2\pi$

$$3h^2 = R^2$$

$$h = R/\sqrt{3}.$$

This is the required relationship! To find the actual volume, we have only to substitute for $h$ in the equation

$$V = 2\pi h R^2 - 2\pi h^3,$$

so that,

$$V = \frac{2\pi R^3}{\sqrt{3}} - \frac{2\pi R^3}{3\sqrt{3}}.$$

Now multiply the first term by 3/3:

$$V = \frac{6\pi R^3}{3\sqrt{3}} - \frac{2\pi R^3}{3\sqrt{3}} \qquad \texttt{Perform the subtraction}$$

$$V = \frac{4\pi R^3}{3\sqrt{3}}. \qquad \texttt{This is the required volume of the cylinder!}$$

We may now compare the volume with that of the sphere:

$$\frac{\text{Volume of sphere}}{\text{Volume of cylinder}} = \frac{\frac{4\pi R^3}{3}}{\frac{4\pi R^3}{3\sqrt{3}}} = \sqrt{3} \cong 1.73,$$

which shows that the volume of the sphere is nearly double that of the largest inscribed cylinder!

PROBLEM 6 (see Page 51)
There are three towns (see Figure 40) we may call A, B and C, whose distances make up an isosceles triangle (the two sides are equal). Town A is in the north, distances AB and AC are the sides, etc. Now the distance from town A to the base BC is 20 miles, while the distance between towns B and C is 30 miles. Find the distance of the point P, south of town A, such that the sum of the distances to all the 3 towns, A, B, and C, to the point P, shall be a minimum, and find this minimum distance to the 3 towns.

SOLUTION
Let us locate the point P which we will find such that the sum of the distances from it to towns A, B, and C will be a minimum. We can let distance AP be either $x$, or $20 - x$. We note that

we will have to square the bottom piece, which makes up two essential right triangles with towns B and C. So we call the bottom distance $x$ for simplicity. (We hope it will turn out that way. There is no guarantee it will!) So the distance between P and B we call $d_{PB}$. It will be (by the Pythagorean theorem)

$$(d_{PB})^2 = x^2 + (15)^2 \quad \text{or} \quad d_{PB} = \pm\sqrt{x^2 + 225}.$$

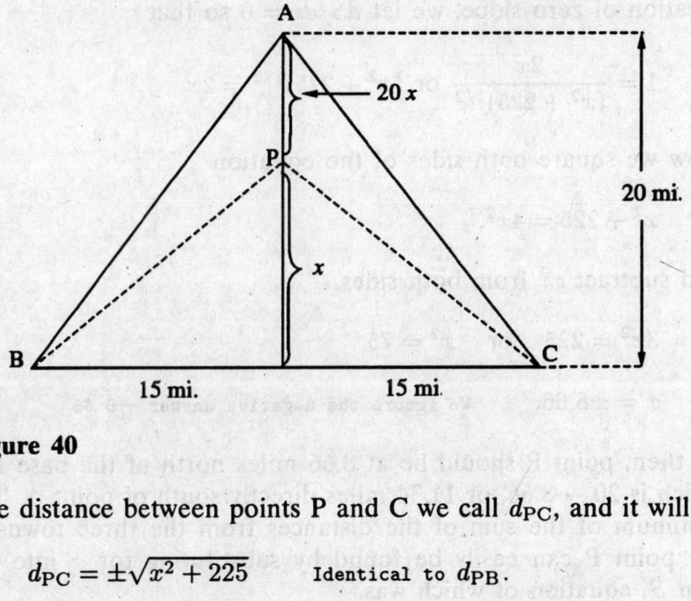


**Figure 40**

The distance between points P and C we call $d_{PC}$, and it will be

$$d_{PC} = \pm\sqrt{x^2 + 225} \qquad \text{Identical to } d_{PB}.$$

Then the sum of all three distances we could call $S$, so that

$$S = \text{dist } AP + d_{PB} + d_{PC}$$

or $\quad S = (20 - x) + 2(x^2 + 225)^{1/2}.$

We have $S$, the sum of the distances, which is the thing we are looking for, alone on the left-hand side. While there are no other unknowns except $x$ on the right-hand side of the equation, we differentiate (this is the easiest part of the problem, as you can see. I might mention that *formulating* real-life problems is probably the most difficult thing). We use form 6 of Table 1.1, which is

$$\frac{d(V^n)}{dx} = nV^{n-1}\frac{dV}{dx},$$

then, $$\frac{dS}{dx} = -1 + (x^2 + 225)^{-1/2}(2x).$$

Since we are interested in the minimum value of $S$, that is the location of zero slope, we let $dS/dx = 0$ so that

$$1 = \frac{2x}{(x^2 + 225)^{1/2}} \text{ or } (x^2 + 225)^{1/2} = 2x.$$

Now we square both sides of the equation

$$x^2 + 225 = 4x^2,$$

and subtract $x^2$ from both sides.

$$3x^2 = 225 \quad \text{or} \quad x^2 = 75$$

$x = \pm 8.66.$ `We ignore the negative answer −8.66`

So then, point P should be at 8.66 miles north of the base BC, which is 20 − 8.66, or 11.34 miles directly south of point A. The minimum of the sum of the distances from the three towns to the point P can easily be found by substituting for $x$ into the sum $S$, equation of which was,

$$S = (20 - x) + 2(x^2 + 255)^{1/2}$$

$$= (20 - 8.66) + 2[(8.66)^2 + 225]^{1/2},$$

calculating which,

$$S = 11.34 + 34.64 = 45.98 \text{ miles.}$$

`(The minimum sum of the distances and the answer)`

Notice that we can do a problem of this kind with 4 towns, or with 15 towns since the principle is the same! The only trouble is that we run into severe algebra difficulties after we let $dS/dx = 0$

and try to solve for a great number of points. For more than 4 points it would be wise to use a computer.

The computer program for this problem is very simple, and even 100 towns (or points) can be tackled in no time at all. But by slope-finding, forget it! (See Program 1 in Chapter 1 for a computer-type solution. Insert the entire new equation in Step 40. If the equation is very large, use several steps.)

### Problem 7

There is a 50 boat fishing fleet, each boat of which can catch an average of 5,000 lbs. of fish. For each additional boat there is a corresponding drop in the catch of 50 pounds of herring per boat. Find the number of boats that should go out so that the entire fleet can catch the largest amount of fish. What is that amount of fish?

### Solution

First we let $P$ equal the total catch of the fleet. Now $P =$ (catch per boat) × (number of boats). Let the number of boats above 50 be called $n$. Then the total number of boats must be $50+n$. And if each boat caught 5,000 lbs. of fish, but for each new boat the loss is 50, then the catch per boat must be $5{,}000-50n$. Then the total catch $P$, must be related in this way:

$$P = \underbrace{(50+n)}_{\text{Number of boats new and old}} \underbrace{(5000-50n)}_{\text{Catch per boat}} .$$

We have the equation already in almost exactly the form we want it. We'll just multiply it out to make it simpler,

$$P = 250{,}000+5000n-2500n-50n^2 = 250{,}000+2500n-50n^2.$$

Now we'll use the slope-finding method since we have the thing we want, $P$ versus $n$, the only variable. Now differentiating,

$$\frac{dP}{dn} = 2500 - 100n. \quad \text{(See Form 3 of Table 1.1)}$$

Letting $\dfrac{dP}{dn} = 0.$ Because we are only interested in the location of $n$ at zero slope

$$100n = 2500.$$

Therefore, $n = 25.$ New boats to be added to the 50.

This means of course, that the fishing fleet should have a total of 75 ships for a maximum catch. And this maximum catch is obtained by substituting for $n$,

$$P = (50 + n)(5000 - 50n)$$

$$= \overbrace{(50 + 25)}^{\text{Number of boats new and old}} \quad \overbrace{[5000 - (25)(50)]}^{\text{Catch per boat}}$$

$$= (75)(3750) = 281{,}250 \text{ lbs. of fish}$$

as the total maximum possible yield. Now for demonstration purposes, a little more realism, we also want to include the cost of manpower, whose work can be put in terms of the value in pounds of fish (for simplicity's sake). Let us say that for each boat over 50, the salary per boat is equivalent to 500 lbs. of fish (the problem could also have been entirely done in terms of money as well). Then the cost is $500n$ for additional *salaries.* Likewise, the equivalent cost for additional *gasoline* is $50n$. Thus the yield must still be equal to the catch less the costs, which in this case is *salary* and fuel. The equation now becomes

$$P = \overbrace{(50 + n)(5000 - 50n)}^{\text{Fish catch}} - \overbrace{500n - 50n}^{\text{Additional costs}}.$$

Simplifying it, we get

$$P = 250{,}000 + 2500n - 50n^2 - 550n$$

$$= 250{,}000 + 1950n - 50n^2.$$

Now we are ready to find the slope; we differentiate

$$\frac{dP}{dn} = 1950 - 100n.$$

Let $\dfrac{dP}{dn} = 0,$

$$100n = 1950$$

$$n = 19.5.$$

We have for the maximum yield either 19 or 20 additional boats, since we cannot send out half a boat. If you really have to know which answer is better, 19 boats or 20, we can simply substitute each of them and see which figure is actually better in terms of maximum yield $P$.

The new total yield with a fleet of 70 ships (i.e., 20 boats added) would now be,

$$P = 250{,}000 + 1950(20) - 50(20)^2$$

$$= 269{,}000 \text{ lbs. of fish.}$$

With a fleet of 69 ships (i.e., 19 ships added),

$$P = 250{,}000 + 1950(19) - 50(19)^2$$

$$= 269{,}000 \text{ lbs. of fish.}$$

Since both answers produce the same result, we can go with 19 boats. Any more boats, for these conditions, is a waste.

So you see, it doesn't matter how many new factors are added, as long as you can turn them into symbols and enter them into the equation. These methods which are shown do actually work for real problems (and are really used). They are not just theoretical questions.

# Appendix B
## Why Division by Zero is not Allowed

| | |
|---|---|
| Let $a = b$. | We will now treat each side equally in all the operations following |
| $a^2 = ab$. | We multiply by $a$ |
| $a^2 - b^2 = ab - b^2$. | Subtract $b^2$ |
| $(a-b)(a+b) = b(a-b)$. | Factoring |
| $a + b = b$. | Divide by $(a-b)$ |
| $b = a + b$. | Reversing the equality |
| $a = b + b$. | Replace $a$ by $b$ in the right-hand side and $b$ by $a$ in left-hand side |
| $a = 2b$. | Add $b + b$ |

This will not do, and makes for chaos in mathematics. We started by saying $a = b$, now we have $a = 2b$. You can see there's no use using the correct rules in mathematics if we use even one incorrect rule. The one incorrect rule in the above was dividing by zero $(a - b)$. Since we specified $a = b$, then $a - b = 0$. We have divided by $a - b$, because we didn't notice it. We cannot (even inadvertently) divide by zero, or by its equivalent.

You can divide by anything else, even a number so small that it's almost zero. You should know that you will get, unsuspecting, even more weird answers to problems when you don't realize you've divided by zero, especially on the computer or the calculator. On occasion, these machines do not realize they have divided by zero, and the answer, though incorrect, might even seem plausible. Therefore take special pains to pore through your equations, especially denominators, to examine them against dividing by zero at any stage.

# Appendix C
# Computer Solution

## Pipeline problem (Page 46)

The program uses a "brute force" solution, i.e., the computer does all the work and "hunts out" all the combinations.

These results follow for the increment of $x = 0.01$ (we truncate at 7 digits).

| $x$ | $P$ | |
|---|---|---|
| 0.96 | 15.81167 | |
| 0.97 | 15.81154 | |
| 0.98 | 15.81145 | |
| 0.99 | 15.81140 | |
| 1.00 | 15.81138 | ⟵ Always the minimum |
| 1.01 | 15.81140 | value for $P$ is at $x = 1.00$ |
| 1.02 | 15.81145 | |
| 1.03 | 15.81154 | |

To satisfy yourself that $P$ is a minimum at $x = 1.0$, try increments of 0.001, instead of 0.01, then change the program in the following way to inspect a yet smaller interval.

| | | |
|---|---|---|
| 10 | x = 0.99 | You might as well start closer to 1 |
| 40 | x = x + 0.001 | The "jump", or increment is changed to 0.001 |
| 50 | IF x > 1.01 GO TO 100 | Shorten the test |

All the other steps are identical. You may try as small an increment as you wish—0.0001 or 0.00001—but the results will be the same, i.e., the value for $P$ will definitely be a minimum exactly at $x = 1.0$. Of course you can only go so far as your computer's ability, i.e., the maximum number of digits it can use.

# Appendix D

Explanation of how $dy/dx = nx^{n-1}$ comes from the function $y = x^n$, and proving it for $n$ values.

From previous pages we see that we needed to find $y_2$ for the slope formula, and for the form $y = x^2$, we needed $(x + \Delta x)^2$. This we can show in the following way. (Watch the exponent.)

$$\text{If } (x + \Delta x)^2 = (x + \Delta x)(x + \Delta x) = x^2 + 2x(\Delta x) + \Delta x^2,$$

and if we need to find $(x + \Delta x)^3$, we would multiply it out to this:

$$\begin{aligned}(x + \Delta x)^3 &= (x + \Delta x)(x + \Delta x)(x + \Delta x)\\ &= x^3 + 3x^2(\Delta x) + 3x(\Delta x)^2 + (\Delta x)^3.\end{aligned}$$

Notice that the coefficient of the $x^n$ terms, when multiplied by its exponent and divided by the number of the term gives the coefficient of the next term. Thus if we multiply the coefficient 3 of the second term, by the exponent 2, and divide by the number of the term, 2, we get 3, which will be the coefficient of the next term, and so on.

And if we need to find $(x = \Delta x)^n$, we follow the same multiplication rules (the Binomial Theorem[1]).

$$\begin{aligned}(x + \Delta x)^n &= \overbrace{(x + \Delta x)(x + \Delta x)......(x + \Delta x)}^{n \text{ times}}\\ &= x^n + nx^{n-1}\Delta x + \frac{n(n-1)x^{n-2}}{2}\Delta x^2 + \ldots\end{aligned}$$

(The dotted line means these terms go on and on using the same rules.) Now that we have $(x = \Delta x)^n$, we continue in the same way as in this book to find $\Delta y/\Delta x$.

---

[1] See any High School book on algebra, e.g. Dressler & Rich–Eleventh Year Math (Ansco), p. 401.

From pages before, we know

$$\frac{dy}{dx} = \frac{\Delta y}{\Delta x}_{\Delta x \to 0} = \frac{y_2 - y_1}{x_2 - x_1} = \frac{(x+\Delta x)^n - x^n}{\Delta x}.$$

Now looking at the last term

$$\frac{dy}{dx} = \frac{(x+\Delta x)^n - x^n}{\Delta x},$$

and substituting for $(x+\Delta x)^n$, we have

$$\frac{dy}{dx} = \frac{x^n + nx^{n-1}\Delta x + \frac{n(n-1)}{2}x^{n-2}\Delta x)^2 + \ldots - x^n}{\Delta x}.$$

Notice we can eliminate the first and last terms, $x^n$;

$$\frac{dy}{dx} = \frac{nx^{n-1}(\Delta x) + \frac{n(n-1)}{2}x^{n-2}\Delta x^2 + \ldots}{\Delta x}.$$

Each term following will increase the exponent of $\Delta x^2$ by one

Now divide all terms through by $\Delta x$;

$$\frac{dy}{dx} = nx^{n-1} + \frac{n(n-1)}{2}x^{n-2}\Delta x + \ldots$$

Now, as once before, we drop all terms with $\Delta x$ since $\Delta x$ will no longer be of any influence compared to the first term. Certainly $\Delta x^2$ is much much smaller than $\Delta x$, and so is $\Delta x^3$, etc. Try it. Start with $\Delta x$ = a very small number, and see what $\Delta x^2$ is. We are now left with the identity we wished to prove:

$$\frac{dy}{dx} = nx^{n-1}.$$ For all values of $n$ (except $n=1$; which will give a zero exponent to multiply with)

# Appendix E

A numerical integration (area-finding) program for the computer (in BASIC). "Rem" means "Remark", and is not acted upon by the computer. We are integrating $y = x^2$ (step 40), with boundaries $x = 0$ and $x = 2$ (step 200).

PROGRAM

```
10   REM*Simpson's Rule
20   REM*Line 40 defines the function
30   REM*Line 200 gives the boundaries of
     integration, the left one first.
40   DEF FN S(X) = X^2
50   READ F, T
60   PRINT "# of increments";
70   INPUT I
80   LET W = (T - F) / I
90   LET A1 = FN S(F)
100  FOR C = 1 TO I - 1
110  LET X = F + C * W
120  IF C/2 = INT(C/2) THEN 150
130  LET A1 = A1 + 4*FN S(X)
140  GOTO 160
150  LET A1 = A1 + 2*FN S(X)
160  NEXT C
170  LET A1 = A1 + FN S(T)
180  LET A = A1 * (W/3)
190  PRINT "Area is: "; A
200  DATA 0, 2   Data 0, 2 are the boundaries of integration
210  END
```

**This program and that in Appendix F will run on the Apple* computer. Changes may be required for other systems.**

Run

```
# of increments? 10  (You must choose an even number)
Area is: 2.66666667  (The exact analytical answer: 2 2/3)
```

*Instructions on its use:* For functions other than the above – only 2 lines need be changed, lines 40, 200. Line 40 defines the function *in terms of X.* (For example, if you wish to integrate

$$G = 4z^2 - 3z,$$

you must change line 40 to read `DEF FN S(X) = 4 * X∧2 −3 * X`. Notice $z$ was changed to X.) In BASIC, the symbol * is for multiplication, ∧ is for exponentiation. Line 200 defines the boundaries of integration (area-finding) with the left boundary given first, the right boundary second. Merely change 0, 2 to the boundaries you are using.

How many slices (increments) to use? This depends on the accuracy you want and the time it takes. Try different amounts of slices, i.e. try $n = 20, 40, 100$. If the answer given does not change much in accuracy with each value of $n$, then that value of $n$ you are using is sufficient. For this method (Simpson's), you must use an even number of pieces ($n$).

In the above example where we evaluate

$$\int_0^2 x^2 dx,$$

$n = 2$ slices and $n = 100$ slices gives the same accuracy to 8 places. But with other functions, for example

$$\int_0^5 e^x dx,$$

this will not be the case, and more slices will be necessary for great accuracy, if that is needed. On the computer in front of me, $n = 100$ slices took 7 seconds for the integrand $x^2$. Of course every additional operation in the integrand will take more time.

# Appendix F

A computer program for the numerical value of the derivative (in BASIC).

**PROGRAM**

```
5    PRINT TAB(10) "x = 1111 stops the program.
     THEN 'List 40' and insert your equation."
10   PRINT TAB(10) "Trig. Functions are in Radians"
20   PRINT TAB(10) "Where y is infinite, the
     slope will be inaccurate."
30   PRINT
40   DEF FN C(X) = X ∧ 2                  We test the function
50   INPUT "Derivative at X = "; X1       Y = X^2 at step 40
60   IF X1 = 1111 THEN 150
70   D = 0
80   FOR N = 1 TO 10
90   D1 = D
100  X = X1 + .5 ∧ N
110  D =(FN C(X) - FN C(X1))/(X - X1)
120  Next N
130  PRINT TAB (24) "is "; 2 * D - D1
140  GOTO 50
150  END
```

**We find the derivative for various values of X**

```
Run
X = 1111 stops the program. Then 'List 40' and insert
your equation. Trignometric functions are in radians. Nearby
to where Y is infinite, the slope will be inaccurate.

Derivative at X = 1       is 2 (Computer answer)
Derivative at X = 2       is 4.00000381
Derivative at X = 8       is 16.0000458
Derivative at X = −25     is −49.9998851
Derivative at X = 1111
```

# Appendix G

Why the constant $C$ cancels out in the definite integral (the boundaries are known).

If we let the $C$ remain after integrating $x^2$ (for example), given the limits $a$ and $b$,

$$A = \int_b^a x^2(dx) = \left[\frac{x^3}{3} + C\right]_b^a .$$

Now we first substitute "$a$" for "$x$", then "$b$" for "$x$". Then we subtract the "$b$" terms from the "$a$" terms;

$$A = \left[\frac{a^3}{3} + C\right] - \left[\frac{b^3}{3} + C\right]$$

The negative sign applies to both of the latter terms in the parentheses

$$A = \frac{a^3}{3} + C - \frac{b^3}{3} - C$$

The $C$ always cancels out, no matter what the integral or the limits

$$A = \frac{a^3}{3} - \frac{b^3}{3}$$

$$A = \frac{a^3 - b^3}{3}.$$

Conversely, if the boundaries are unknown, then the $C$ must remain. The shape of the curve is determined by the other terms, but the $C$ calibrates it, i.e., it shows where the curve belongs.

# Bibliography


1. Silvanus Thomson, *Calculus Made Easy*, St Martin's Press, New York, 1988.

2. Simeon Berman, *Calculus for the Non Physical Sciences*, Holt, Rinehart and Winston Inc., New York, 1974.

3. George B Thomas, Jr., *Calculus and Analytical Geometry*, Addison-Wesley Publishing Co., Readfing Mass, 1996.

4. I N Bronshtein and K A Semendyayev, *A Guide Book for Technologists and Engineers*, Pergamon Press, Oxford, London, 1997.

5. Earl K Bowen, *Mathematics with Applications in Management and Economics*, Richard Irwin, Inc., Homewood, Illinois 1976.

6. Harold R Jacobs, Mathematics, *A Human Endeavor*, W H Freeman and Company, San Francisco, 1982.

# Index